DARWIN AND THE MONSTERS

Regal tells the story of the engagement of philosophers, naturalists, and scientists in what we now call "cryptozoology," a much longer history than most appreciate.

Most believe cryptozoology – the study of mysterious, unrecognized creatures – emerged in the 20th century. But the fascination with monsters and the unknown stretches back to the ancient world. Long before the term was coined, philosophers and naturalists pondered the existence of cryptids, weaving them into the fabric of scientific curiosity. Challenging the notion that mainstream science has always dismissed cryptids, it reveals a rich history of inquiry and debate.

Offering a groundbreaking examination of the long and complex history of cryptozoology, Regal situates it within broader debates about biology, evolution, and the limits of scientific knowledge. A fascinating read for students and scholars of cryptozoology, the history of biology and zoology, and science, technology, and society studies.

Brian Regal is Professor of the History of Science, Technology, and Medicine at Kean University, USA.

DARWIN AND THE MONSTERS

The Long History of Cryptozoology

Brian Regal

Routledge
Taylor & Francis Group
LONDON AND NEW YORK

Designed cover image: Figuier, Louis, and W. S. O. *The Ocean World: Being a Descriptive History of the Sea and Its Living Inhabitants.* London: Chapman & Hall, 1868

First published 2026
by Routledge
4 Park Square, Milton Park, Abingdon, Oxon OX14 4RN

and by Routledge
605 Third Avenue, New York, NY 10158

Routledge is an imprint of the Taylor & Francis Group, an informa business

British Library Cataloguing-in-Publication Data
A catalogue record for this book is available from the British Library

ISBN: 978-1-032-78634-6 (hbk)
ISBN: 978-1-032-78633-9 (pbk)
ISBN: 978-1-003-48876-7 (ebk)

DOI: 10.4324/9781003488767

Typeset in Sabon
by Apex CoVantage, LLC

CONTENTS

ACKNOWLEDGMENTS

Thanks to . . .

C. T. Ambrose, B. J. Gooch, and Charles Boewe, for their help on Constantine Rafinesque

Student researcher Renee Hertz

The staff at the Bodleian Library, Oxford, and the library of the British Museum of Natural History, I'd like to thank the librarians at the Natural History Museum, London, for their help with the Owen papers.

Mrs. Jane Bradley, Local Studies Librarian, Bristol Reference Library, Bristol, UK

Martin Graebe of the Sabine Baring-Gould Appreciation Society

Dr. Joanne Parker, University of Exeter

Claire Hopkins, archivist at Trinity College, Oxford

Caroline Palmer of the print room, Ashmolean Museum, Oxford

INTRODUCTION

The world is full of monsters. We try to avoid them, being afraid of what they might say to us, or about us. And sometimes we try to find them and make sense of them. There are many types of monsters. There are psychological monsters and physical monsters: creatures of the mind, and those of nature. We often label a thing or a person a monster because we do not understand them. We fear most that which we do not understand. A few brave souls chase them to see if they are real.

Physical monsters from the natural world tend not to survive too long. They are seen as monsters until it is shown that they are fish, or birds, or primates, or things that simply do not exist. Those people we think of as scientists – philosophers, academics, learned persons – are usually thought to be too grounded in reality, or too skeptical to believe. As a result, there is a long tradition of a gulf between the believers and skeptics in physical monsters. This book is an attempt to show there is an equally long history of scholarly interest in creatures we call monsters, but not as ethereal objects of philosophical inquiry, but as living, flesh-and-blood creatures.

Mr. Darwin

Sitting in his comfortable study in 1844, an English naturalist sat pondering and perusing a book he had recently acquired. It was a book about monsters: more accurately, a book about monstrous births of both humans and animals. Such occurrences as two-headed babies, babies with no limbs or an excess of limbs, conjoined twins, too many eyes or not enough eyes (today known as birth defects), fascinated naturalists, philosophers, and the public alike. These types of births stirred both fascination and revulsion in

DOI: 10.4324/9781003488767-1

people for a long time. The English word "monster" derives from the Latin *Monstrum,* which meant a portent or prodigy. Before the modern advent of genetics and heredity many people thought the appearance of a deformed animal or human infant presaged some terrible disaster. Their causes were put down to supernatural origins rather than naturalistic ones. Because such births were seen as ugly and against the norm, the term came to eventually mean a horrid and even evil creature.

As he turned the pages, the English naturalist made marginal notes in pencil, he thought might help him with the problems he encountered in a research project he was working on. The book, written by French naturalist Isidore Geoffroy Saint-Hilaire was *Histoire générale et particulière des anomalies de l'organition chez l'homme et les animaux,* better known as the "Treatise on Teratology" delighted the reader. Saint-Hilaire's text was a detailed study of monstrous births. He tried to explain them from the standpoint of biology rather than theology.

The English naturalist thought that monsters such as Saint-Hilaire (1772–1844) wrote about might shed light on the growing vexed question of biological diversity, how generation worked, and possibly even how living things transmuted from one form to another. The naturalist's name was Charles Darwin. The work of Isidore Geoffroy Saint-Hilaire was going into the grand soup that would eventually be published some years later as Darwin's groundbreaking book *On the Origin of Species* (1859).

There is a widely held belief today that mainstream science and history have always rejected monsters as fables and hoaxes, that they are the products of fantasy, of minds rejecting rationality and cold hard science in favor of the romantic. This is simply incorrect. This is a tradition that goes all the way back to the ancient world and right up to the 21st century. Monsters have always occupied a strange, unsettling, and dark place in the history of science in general and the history of evolution in particular.[1]

There is an enormous body of literature on monsters and what they mean to humans. It reaches back to the ancient world and crosses all cultural boundaries. There is an equally impressive modern body of work on the meaning and place of monsters in human folklore and psychology, which shows how these creatures entered the human psyche, what they do there, and what we use them for. This research shows the metaphorical and allegorical place of monsters in human history and has been labeled by author Jeffrey Cohen as "monster theory."[2] Throughout this entire history, some authors, however, pursued monsters as genuine biological organisms. It is this aspect of the story I plan to examine in this book.

Original texts on monsters are often read as allegories, metaphors, and as vehicles to examine other things such as race and gender, clashes of cultures, questions of empire, the darker and deeper parts of the human psyche, and others. This approach has yielded much useful knowledge and understanding of the human condition. It is taken as a given that since we know today that our view of what monsters are shows they do not exist nor ever did, that earlier authors also saw monstrous creatures in metaphorical and folkloric terms alone. Modern researchers must be careful of this presentist approach. While historic authors certainly appreciated the ephemeral nature of some monsters, they also saw them as part of the natural world. Believing them to be genuine biological realities in the wider pantheon of living things left naturalists and other commentators no choice but to examine these creatures at face value: to explain their origins, their morphology, their relationship to other living things, and to either accept them as real or show their dubious nature. It is in this effort to come to terms with the physicality of monsters which is the heart of this book. For in the engagement with the notion of monstrous creatures, these authors helped pave the way for the modern understanding of biological diversity, evolution, and medicine. By the 19th century, belief in the genuine nature of monsters had waned, but what might be called the language of monsters had become so ingrained into biological thinking that explanations of life on earth still reeked of it. Monsters played a subtle role in the great debate on Darwinian evolution and were used by both pro and anti-evolution proponents. (Indeed, in the 21st century, anti-evolutionists still employ monsters as a way of discrediting evolution).

This book, however, will eschew this "monster theory" and look to something else. I will attempt to trace the hunt for and engagement with monsters in their physical form: as products of the natural world rather than simply of the human mind. It will examine how the hunt for and attempt to explain monstrous creatures in all their guises contributed to the modern understanding of biological classification and evolution theory. Evolution would help put the final nail in the coffin of the sciopods and werewolves. Yet, in a fascinating turnaround, evolutionary theory would, in the later 20th and early 21st centuries, lead to a renewed interest in monstrous creatures that would take the form of the discipline of cryptozoology.

The text here focuses on the Western monster experience. All nations and cultures have monsters as part of their cosmologies. All world religions also talk about monsters in various ways doing various things. They are mostly seen as metaphors used for addressing questions of metaphysics such as the afterlife and spirituality. There are few examples of people from these cultures searching for monsters as flesh and blood organisms, however. The fascination with proving monsters real tends to be a Western European preoccupation. When monstrous creatures in Africa, Asia or Latin America are searched for, the search is conducted by Euro-Americans, not indigenous locals.

Monsters, both real and imagined, could adroitly hopscotch across the boundaries of philosophy, science, religion, and literature. Monsters had the agility to be in several places at once. All the time, "science" and scientists frantically tried to get control over such creatures. Sometimes they were successful at corralling monstrosity, other times not. The scientific impulse pushed naturalists to look for factual evidence rather than relying on anecdotes and oft-told stories. Despite episodes where scientists proved themselves as vulnerable to nonsense as anyone, they soon pushed back and sought the underlining reality or hoax nature of the reports.

So, What Is a Monster?

There are a number of definitions used to describe a monster. It can be a loaded and ambiguous question. Originally, it meant something which brought along a portent of the future. In that way, a monster could be anything from the death of a King, to the birth of a two-headed baby, a strange astronomical event in the sky, or the arrival of a group of bandits. It took on its modern meaning – a hideous and dangerous creature – when applied to the birth of a deformed child or even animal. Not understanding the biological mechanics of birth defects or genetics, early people sought the only recourse they had: they ascribed such monstrosities to divine or demonic interference. As a deformed child or animal was ugly and went against the normal the meaning of monster began to shift. Also, as the concept of portents of catastrophes, failed crops, and general bad times began to lessen (though never really disappear) monster became a term to describe people or animals physically ugly and deformed. The portentous aspect took a back seat to simple, physical ugliness. As ugliness was soon equated with dangerousness the modern meaning was born.

In this study, I'm going to cast the definition of monster somewhat narrowly. I will not include monstrous people such as murderers, corrupt politicians, religious cult leaders or dictators, for example. I will also not include fictional monsters from literature and film. I will focus on monstrous creatures that transcend the mythical and are thought, by some, to be living biological entities that have yet to be scientifically described and accepted as such.

There is ample literature on the historical development of various fields of scientific inquiry. The transition of geology, ornithology, botany and many others, from amateur pastimes to legitimate, respected, professional careers is well documented. I will look at the reverse, however, to the de-legitimization of a field. How does a field go from professional to amateur, from embraced to abandoned? How did the study of monsters go from a respected field to that of an inconsequential, even suspect, form of inquiry?

Historiography

There is a strange paradox in monster studies. Amateur enthusiasts write extensively about Bigfoot, the Chupacabras, Jersey Devil, and other such entities we might classify as modern pop culture creatures: ones that fall under the general rubric of "cryptozoology." These creatures often find themselves left out of academic discussions of monsters. It seems that scholarly, historical examinations and serious philosophical studies of, say, scipodii, are deemed academically acceptable, even the subject of high-end peer review books and doctoral dissertations, while studying Bigfoot is relegated to unconventional and outsider authors. Both, however, are products of popular culture. As Lorraine Daston and Katherine Park argue in their excellent *Wonders and the Order of Nature* (1998), classical monsters all had their origins in contemporary pop culture. In this way, the Blemmyae are no different from the Yeti.[3] In the academic world it seems, no monsters are allowed to pass the 18th century. Amateur monster authors have produced libraries full of works on the Beast of Bodmin, the Snallygaster, Opobowa, and others, yet rarely if ever mention the classical monsters. These authors seem as uncomfortable talking about the Scipodii as the academics feel about Bigfoot. Neither side likes to cross the 18th-century monster barrier, though they are all part of the same continuum.

The modern pursuit of monsters is called cryptozoology, a term which first appeared in the post-WWII era. Its antecedents, however, go back to the earliest days of the human experience, and likely back to our proto-human ancestors. Helping this concept along was the rise of evolution studies, without which there would be no modern notion of cryptozoology – despite the number of monster hunters who self-identify as creationists or attempt to use cryptozoology to discredit evolution.

Twentieth-century practitioners of cryptozoology argue that what they do is a search for hidden animals in a proper, empirically based science rather than a paranormal enterprise as some detractors argue. The search for monstrous creatures – and the official birth of cryptozoology – in the 1950s was led primarily by trained scientists rather than amateur monster hunters, though amateurs certainly contributed to the effort. The reasons 20th-century academics took on monsters came not through a dreamy sense of romance of far-off places and mysterious animals, but because their training in science and evolution studies told them if any strange organism, no matter how improbable sounding, existed it did so because of the same evolutionary mechanics as any other organism. So, if the Loch Ness monster or Bigfoot lived, it was the result of millions of years of natural selection, not magic or the supernatural.

A long line of authors, naturalists and scientists contributed unknowingly to the emergence of cryptozoology. Along with the ancient authors such as Aristotle, Pliny, and Lucretius more modern researchers such as A.C. Oudemans, Constantine Rafinesque, Richard Owen, T.H. Huxley as well as the great systematizer Carolus Linnaeus and even Charles Darwin was influenced by monsters and used monstrous imagery and lexicon in their work. It must be said, clearly and unambiguously at this point, however, that Charles Darwin in particular, and these others generally, were not cryptozoologists. The argument I am trying to make is that monsters – whether animal or human and in their many guises and meanings – played an important but underappreciated role in the establishment of the sciences of biology, paleontology, and evolution studies, and that they in turn played an important role in the creation of a new version of monster studies when these monsters were eventually abandoned by the mainstream. Evolutionary science helped discredit the notion that monsters were the result of a mating between Gods and humans or some other supernatural or magical cause. It helped banish monsters from mainstream scientific thinking, but that vacuum soon filled with those who engaged in cryptozoology.

Like many scientific endeavors, monster hunting mirrored wider social and cultural conditions. As with any "search" humans engage in, the search for monsters is a search for the self. It matters little whether the object searched for exists as a physical reality. Like Lewis Carroll's fictional hunting of the Snark, the object is barely known to the searchers, yet they search for it anyway.[4] Even if never found, such a search can lead to new discoveries and insights about life, nature, or ourselves. The age-old search for monsters has been about more than just running down a beast. It has been about the place humans hold in the great diversity of life, it has been about where we come from, what it means to be human. It has been about our relationship to the natural world, about our relationship to each other, about the value of nature. Monsters, it seems, have a special ability to help us examine these things.

One underappreciated aspect of monster studies is that the search is less about flesh and blood creatures than it is an engagement with the history of human hatred and fear. It is common to all human cultures to take what we don't understand, that which we fear, and turn it into a monster. We monsterize those things we don't like, from people of different ethnic backgrounds to different genders to different opinions. The first step along the road to genocide, for example, is to turn the people we don't like into monsters. That way it is easier to shovel them into ovens. Our monsters tell us more about ourselves than about any ill-defined "them." This book is an examination of how science approached these creatures.

Having said that, it must be noted that the word "scientist" is problematic for a story that covers so much chronological ground. We cannot accurately use the word to describe people from the ancient world as scientists in our modern sense of the word. The same holds for the medieval and even into the modern period. The term "scientist" was not coined until the early 1830s when British historian William Whewell began using it in his writing.[5] I am left with this conundrum of terminology. In an effort to circumvent confusion – and purists who would argue Lucretius, for example, was not a "scientist" – I will use the term "learned person" and even scholar, though I do it with the understanding that it might be applied incorrectly to some individuals. I hope this will help the reader to appreciate that I am talking about individuals who studied, in whatever vague way we chose to define, the universe in general and living things in particular, as the object of this work.

These individuals saw monsters as a guide to understand life on Earth. Monsters were learning agents and teaching aides. They saw that monsters scream at us with the raging energy of the now and moan with the desolation of forgotten things.

Notes

1 Lorenzo Rossi and Veronica Padovani, "The Science of Monster Hunting," *Atti della Società dei Naturalisti e Matematici di Modena* 152 (October 2021).
2 Jeffrey Jerome Cohen, *Monster Theory: Reading Culture* (University of Minnesota Press, 1996). Also see, Steven Asma, *On Monsters: An Unnatural History of Our Worst Fears* (New York: Oxford University Press, 2009). M Susanne Schotanus, *Interdisciplinary Essays on Monsters and the Monstrous* (Bingley, UK: Emerald Publishing, 2022).
3 Lorraine Daston and Katherine Park, *Wonders and the Order of Nature* (New York: Zone Books, 1998). Also see, Laura Lunger Knoppers and Joan B. Landes, *Monstrous Bodies/Political Monstrosities in Early Modern Europe* (Cornel University Press, 2004).
4 Lewis Carroll, *The Hunting of the Snark: An Agony, in Eight Fits* (Macmillian, 1876).
5 Sydney Ross, "Scientist: The Story of a Word," *Annals of Science* 18, no. 2 (1962): 65–85.

1

MONSTERS OF THE ANCIENT AND EARLY MODERN WORLDS

Some of the earliest philosophical discussions of monsters and their physical existence come from the authors of the Classical age. As monster enthusiasts do today, they ran the gamut from believers to skeptics. One thing they shared was the belief that whatever they were, and whether real or not, monsters should be studied. Whether they were figments of the imagination or living flesh and blood entities, they needed to be known. If they were real monsters, they should be studied to see if they were part of the wider run of living things. A common motivation for the study of monsters during this period was that they were considered abnormal. Thus, they could be used as a measuring stick against what was normal. Their abnormality had to be better understood so that normality could be better understood.[1] This fascination carried through to the Medieval and Early Modern era, and in fact continues to this day.

It began with the great philosophers of antiquity in their efforts to understand humans themselves. Monsters became devices to better understand the universe as well as the human mind. In the philosophy of Plato (427–347 BCE) Man had three distinct parts. The spirited, the rational, and the irrational (the monstrous). These parts had to work harmoniously, or catastrophe could result. The rational had to work to tame and control the irrational. This was how perfection could be attained. The notion of perfection and how living things were generated, and why some were different, became a major question for those early philosophers interested in living things and where they came from. Plato's archetype concept acknowledged that all things had some original, abstract, perfect form: the Archetype. Every other version of that thing – whether animal, plant or human – had this perfect abstract form. Thinkers following this idea compared other things against these perceived

DOI: 10.4324/9781003488767-2

archetypes. Ancient authors took monsters seriously and asked important questions about them in part to study archetypes. If perfection was a thing, then so should be imperfection. These imperfect forms were deviations from the archetype. The further from the archetype, the more monstrous. Could perfection exist without imperfection? Monsters upset the order of things. In works like his *Symposium,* Plato argued monsters of the human soul represented imperfection. And thus incompleteness. Monsters had to be tamed in order for Mankind to attain completeness and thus perfection. If so, then what was the nature of imperfection? What did imperfection say about living things and how they came about? "Monstrosity is one of the main conceptional challenges for any philosophical system."[2] If all things had purpose, what was the purpose of monstrosity? Unlike other classical writers, Plato saw monsters less as flesh and blood creatures and more the intangible qualities of human behavior. How did it come about and what was its source? Could one go out and see, and even collect and study, monsters? Or were they, as he argued, philosophical constructs of the human condition?

Plato's student Aristotle (384–322 BCE), on the other hand, argued that monsters were physical beings created by luck or chance of birth and were contrary to nature, thus you could recognize them but couldn't figure them out. If a strange creature was a one-off then it was a monster. If there were a population of them, then they were not a monstrosity. Something is not monstrous if it occurs on a regular basis.[3] Conforming to his concept of there being a hierarchy of perfection, Plato thought monstrosity was part of the lower orders of living things. There must be inferior things in order to have superior things.[4] As they surveyed their surroundings, philosophers of the ancient and medieval world saw monsters everywhere. Unlike Aristotle these later naturalists believed monsters could come in populations. This entire book is an examination of scholars trying to figure out if monsters were real.

The Monstrous Races

The more they learned about the wider world the more Greco Roman, then Western European authors saw monsters everywhere. They were born of ideas coming out of India and Persia. India was of special interest for its varied society, flora and fauna, and monster mythology. It all entered the Western consciousness. When Alexander invaded India in 326 BCE Greek writers gained more access to the subcontinent. The work of Ctesias of Cnidos (5th C BCE) covered the monsters of India to the point where it was so influential that it drove the Western view of India and monsters for centuries.[5] He recounted tales of pygmies, giant ants, Blemmyae (people with the faces in their chests) and the other most popular Medieval monstrous race, the Sciapodes (people with one giant foot).[6] The Blemmyae and Sciapodes became so popular they went on to appear hundreds of times in subsequent illuminated manuscripts.[7] Unfortunately, Ctesias' work does not survive complete but had to be reconstructed by tracking down parts of it in other manuscripts.

In the *Indica* (310 BCE) of Megasthenes (died 390 BCE), an important history of India, there is a discussion of Monopods/Sciapodes.[8] This was used as a source work by a number of authors including Pliny the Elder, Eusebius, and St. Augustine. In his *City of God* Augustine ponders whether, what were now being called, the monstrous races are of human origin. He wondered whether the monstrous races came from Adam or the sons of Noah. "Accordingly, it ought not to seem absurd to us," he says,

> that as in individual races there are monstrous births, so in the whole race there are monstrous races. Wherefore, to conclude this question cautiously and guardedly, either these things which have been told of some races have no existence at all; or if they do exist, they are not human races; or if they are human, they are descended from Adam.[9]

Publius Ovidius Naso (43 BCE–17 CE) better known as Ovid, like so many writers of the classical age took monsters at face value. Even when obviously myths they still had important things to say. Centaurs, for example were indicative of blurred boundaries. Monsters also represent political change. Ovid was born into a well-to-do Roman family. He studied to be a lawyer but gave that up in exchange for the life of a poet and philosopher. His most famous and well-known work, *Metamorphosis,* teems with both literal and figurative monsters. For Ovid, like it did for a number of authors, and still does, monsters served the purpose of being vehicles to explore other, broader issues. Greek mythology supplied creatures such as the Minotaur and Medusa for his consideration. These monsters were for careful study as philosophical constructions used to explain the human condition. He wanted proof of monsters not just stories. They were not thought of as flesh and blood organisms one could go out, look for, and find. Monsters for Ovid represented the porousness of boundaries. Especially between men and women, and animals. They also represent change, indeed, the entire text of the *Metamorphosis* is about change from one state to another. He teeters precariously close to articulating an evolutionary worldview. "All is in flux." He says. "Any shape that is formed is constantly shifting. Time itself flows steadily by in perpetual motion . . . The present turns into the past and the future replaces the present; every moment that passes is new and eternally changing." He says he saw a man change into a werewolf with no traces of the man remaining.[10] All the monsters he discusses represent change.[11]

Much of what Western authors knew, or thought they knew, about monsters came from Indian tradition. Hearing about Centaurs and Satyrs the Greeks, for example, wanted to know whether they constituted actual organisms or just legends. The larger idea of "India" was becoming part of Greek cosmology in part thanks to the work of Herodotus and other writers. As it would for many later cultures it fascinated them. Many fevered fantasies about the subcontinent began to spread throughout Greek culture, and then eventually to the wider Western world.

Herotodus (484–525 BCE) is considered the first "historian." This term cannot be understood as having the same meaning it has today. Unlike modern historians he was not particularly enamored of factually based writing. He did not allow facts to get in the way of a good story or myth. As such we cannot take his *Histories* (430s BCE) at face value.[12] It does, however, give us a look into the world of the Greek states and in particular the Greek encounter with Persia. He lets himself off the hook by saying he is simply repeating what the Libyans told him. He seems to have been fascinated with the odd. People are interesting, he said, because they are odd. Monsters also fascinated him.[13] He also seems, at first to be preoccupied with the idea of kidnapping woman and carrying them off. He tells of the "two-fold creature" being half woman and half snake. Arimaspians had only one eye. There are bald headed men and women who are bald from birth. These baldheaded men say that the mountains near them are populated by men with goat's feet, "though I don't believe it." Even Herodotus had his limits. At the extremes of Scythia there are more one-eyed men who keep Griffins to guard their gold. There were the Akephaloi with no heads and their faces in their chests. Medieval authors would know these creatures as the Blemmyae. In the Ramayana of India, the evil Kabandha is described much like the Blemmyae. Also, the cannibalistic Androphagi who Herodotus says "have the most savage manners of all human beings." The Nueroi occasionally become wolves.[14]

Another major author whose work on monsters would have lasting implication was Pliny the Elder (24–79 CE) and his *Natural History*. A philosopher of wide ranging interests, Pliny wrote a number of texts before he led a rescue mission to the eruption of Mount Vesuvius in the year 79 (an expedition he famously did not return from). He was a soldier-scholar (having studied law) as was typical of Roman men of the Equestrian ranks at the time. He served in first the infantry then the cavalry. His last work, *Natural History* draws upon other writers to tell of all the world's marvels (possibly the first encyclopedia). A good bit of it though, is based upon Pliny's own observations and study.[15] He argued that nature was anything worthy of memory.[16]

Building upon the work of Ctesias, Megasthenes, and Herodotus, Pliny discusses, with barely contained glee, what has become the usual suspects of the monstrous races. They come crowding out of India and Persia in his Book VII. Along with Griffins and Cyclops, we meet more Sciapodes, Cynocephally (dog headed men), and Antipodes (whose feet point in the wrong direction). Regardless of whether Herodotus or Pliny actually saw these beings that were described to them, and whether they repeated the descriptions accurately, their words were taken as accurate by their readers. Herodotus may have been the first to go looking for monsters rather than simply relying on various legends and stories. He says he went to Arabia to find flying reptiles: that is, dragons. He came away convinced they were real. While in "Arabia" he claims to have seen the remains of winged serpents. Herodotus' *Histories* became the basis for many of the monster legends and myths that came down from then through the Middle Ages well into the Modern period. Herodotus said the dog headed Cynocephaly lived in Libia, Ctesias said they lived in India as well. The Cynocephally, Blemmyae, Sciapodes, and other monsters who populate Medieval manuscripts owe their existence to these authors. Although, similar "people" are found in Chinese and Japanese mythology. These are the creatures who would become in the European Middle Ages the Monstrous Races.[17]

Satyrs and Centaurs

The Roman author Lucretius found the discussion of monsters fascinating, but puzzling. There was much talk about Centaurs (half human and half horse), for example, but little in the way of evidence, or even logical thinking. How could a creature be part human and part horse? It made no sense. Lucretius searched his library but could find nothing convincing other than unsubstantiated stories. He needed more. He, along with other authors and learned persons had heard the stories and legends attributed to Indian sources, but nothing convinced him. If such creatures existed, they had to be logical, they had to fit into the larger scheme of life on Earth somehow. If they were real they had to be proven so, and as such studied. If not, then they were figments of the imagination: phantasms worthy only of old wives' tales, not serious study.

There is little known of the life of the Roman writer Titus Lucretius Carus (99–55 BCE). It is easier to pin down what he said and dreamt about monstrous creatures. We have no idea if he traveled to foreign lands and saw Centaurs or giants with his own eyes or if he just heard tales of such creatures from travelers in the way Herodotus had. When confronted with such legends and stories, however, Lucretius was often skeptical of them. He had a certain ability to see the inherent absurdity in fantastic claims. In his most famous work, *De Rarum Natura*, better known as *On the Nature of Things*, he does two things: first he explains the nature of his following of the philosophy of Epicureanism. Adherents to this way of thinking believed that life was the pursuit of virtue by seeking modest and sustainable pleasure. It was also a basically skeptical approach to knowledge. Skepticism led to a clearer understanding of the natural world and thus to peace of mind.[18]

He then addresses monsters and fantastic creatures and searches for some insight into why people believe in them. He seems perturbed with the idea that people believed such creatures as dragons and centaurs were real.

Though he never planned it, and did not have a term for it, his interest in creatures vaguely known to us as monsters nudged him to make early steps toward the creation of what would later be known as cryptozoology. At the same time, however, he sowed the seeds of its destruction. "Certainly, the image of a centaur is not derived from a living being since no such creature ever existed."[19] He was equally skeptical about giants. "Why has nature not succeeded in producing human beings so huge they could wade across the open sea, making it seem shallow?"[20] These questions were reasonable and sensible. Not magical.

The Satyrs are an interesting model for the relationship between monsters and scholars. Unlike, say, Sciapodes or Blemeyae, Satyrs may have had a relationship with known, or at least, somewhat known animals. During the Medieval age and early Modern, Western researchers were becoming more aware of what would eventually be called primates. Strange human-like creatures were encountered in Asia and Africa. While knowledge of them was not detailed or scientific, they were known just enough to know they were there. Their existence was much less questioned than other monsters.

Marvels of the East

Along with India, China had a contingent of monsters.[21] There were additionally attempts to understand monsters, spirits, and ghosts. Written by Gan Boa around 350 CE, *In Search of the Supernatural* (*Soushen Ji*) is a collection of short writings and eyewitness accounts of creatures stalking the Chinese hinterlands.[22] Reminiscent of an encyclopedia it articulates the difference among the various dragons appearing in Chinese culture. He popularized the Zhiguai "strange tales" style using eyewitness accounts meant to look like genuine historical fact-based writing. He categorized the many different types of dragons. There was the Jialong (with scales), Yinglong (with wings), Qiulong (with horns), and Chilong (without horns) all of which were distinct entities. Along with the dragons, there are the Fox Spirit, guardian lions, heavenly dog, Jade rabbit, and Zhuyan Ape. It also contains the popular legend of the brave Li Chi who slew a monster serpent that terrorized her village. The monster demands the people give it young girls to eat. Li Chi volunteers to confront the huge creature. She goes to its cave accompanied only by her faithful dog. She baits the monster with sweetened rice balls. When the creature comes out Li Chi cuts off its head with a sword.

Born in Henan Province, Gan Boa (286–336 CE) became fascinated by the supernatural as a young man. He studied Chinese mythology to the point where he was made historian for the Jin Dynasty. Mythology extended to his own life. His father had a year-long affair with the family maid. When he died Gan Boa's mother had the maid entombed alive with her dead husband as punishment. Ten years later when Gan's mother died, they opened the tomb to inter her body. To everyone's amazement the maid was still alive. She said the spirit of the father had brought her food and water to keep her alive. Gan Boa eventually married the maid and had a child with her. There are a number of crucial texts describing monsters in ancient China. Along with *In Search of the Supernatural* there is the *Classic of Mountains and Seas,* and the *Discourses of the States*.[23] In China headless men with their faces in their chests are known as Xingtian, in Japan they are known as Yokai.

Empedocles

One last Greek philosopher that must be mentioned here is Empedocles (494–434 BCE). Though he is considered a pre-Socratic thinker his ideas about how living things came into being, and how to approach monster studies, have a distinct ring of Darwinian evolution to them. As with the writings of Ovid modern investigators see signs of natural selection in the work of Empedocles. Also, like Ovid he uses monsters to make his point. Like Herodotus he may have died in a volcanic eruption: this time Mount Etna. Not much about the life of Empedocles, other than that he was probably born on Sicily, is known for sure.

Empedocles saw various monsters as malformed humans from when the world was young. In the early ages of the Earth the world was a place of great confusion and turmoil. Headless bodies and even disembodied arms and legs wandered the landscape. Some humans had no heads while others had the heads of animals. They did not live long because they were incapable of competing with their non-deformed counterparts. The deformed humans could not adapt to their environment and thus died off. In an early form of natural selection, the deformed humans – the monsters – disappeared while the humans who did adapt survived. Monsters were developmental dead ends. Adaptation in animals and humans was crucial for survival. A person with no head but their face in their chests or with only one giant foot, or the head of a dog was clearly not properly or completely formed. In these states they could not live for very long.

While his explanation for where monsters came from and what happened to them was not well laid out in detail it did say some of the important parts: that better adapted, better formed things would survive and produce offspring while lesser or malformed monsters died off. Though his ideas did not catch on with others, in attempting to explain monsters Empedocles had stumbled upon an early form of evolution theory.

In addition to the philosophical discussions of the ethereal nature of monsters, there were the traveler's tales. These purported to be descriptions of physical encounters with monsters: or at least recounting what others had said about encounters with monsters. A traveler would go to some exotic part of the world – mostly Persia or India – for some other reason but encounter locals who would regale the traveler with amazing stories. Upon their return the traveler would commit his experience to paper. These types of texts brought monsters out of the libraries of learned persons and into the popular consciousness.

Mandeville's Travels

Possibly the most popular early traveler tale of exotic lands came in the form of Marco Polo's *Travels*. Polo (1254–1324), a Venetian trader, went on his famous adventure in the late 1200s. He said he went through India and ultimately to China and the court of Kublai Khan. Marco was following in the footsteps of his father and uncle who had already made the trip some years before. Upon his return Venice and Genoa were at war and Marco fought for the Venetians. He was captured and thrown in a Genoese jail where he met the writer Rustichello da Pisa. Polo began recounting his story and Rustichello wrote it down.

While the text does mention exotic Eastern animals such as crocodiles and rhinoceros there is no explicit mention of the types of monsters of other writers discussed here. The monstrous tradition of the *Travels* come from the fact that illustrated versions of the book include Sciapodes, Blemmyae, and others. This is likely an artifact that by the time these copies of Marco Polo's *Travels* were produced such creatures had become popular subjects for manuscript illuminators.

The other most famous Medieval traveler's tale from this period is that of Sir John Mandeville. Mandeville, who was likely a literary construction rather than a genuine person, is portrayed as a knight who began his travels in 1323 and did not return for 34 years. He was born in England in Saint Alban's and, as a knight, many of his journeys were to various holy lands. He speaks of many different races of humans and animals that would, in modern times, be better referred to as monsters. His description of the different regions is sketchy at best. Historians have debated who the actual author is.[24]

Making a number of sensationalist claims, Mandeville mentions an island with giants twenty-eight feet tall that are cannibalistic; beyond that is another island with 45 to 50 foot giants and sheep the size of oxen. In Ethiopia, there were men with only one foot that move very quickly (the Monopods) presumably by hopping. Their single foot is very large, so that when they lay on their backs it shelters their bodies from the sun and rain. Along with the giant foot creatures there are men with heads of dogs that are "reasonable" and worship an ox-like god. The Blemmyae are men that lack heads and necks but have their eyes on their shoulders and their lips in their breast. Many of the monstrous races made popular by Herodotus, Pliny, and Empedocles roam through the landscape of Mandeville's narrative. He says that bones of men forty feet long were discovered in a place called Jops. Like many of these books, Mandeville's *Travels* is a long adventure narrative in which the author regales the reader with fascinating stories of exotic, far-off lands populated with equally exotic creatures.

Monsters in the Monasteries

Being born in India monsters proceeded to slither, hop, and slouch their way into not only the Western consciousness but into their writings. Mandeville's *Travels*, for example, originally appeared in a French manuscript. It proved so popular that scribes began cranking out editions in various languages. It almost seems that no self-respecting library, whether monastic or private, could be without one. Any text that spoke of monstrous creatures was sought after. In the scriptoria of monasteries across Europe and the Middle East Illuminated Medieval manuscripts became veritable zoos of monsters of every stripe. Some works were concerned primarily with monsters and became known as Bestiaries.

During the Medieval period books were produced in monastery scriptoria by teams of monks working on several projects at once. Before the advent of printing books were produced by hand. There was a division of labor: there were translators, book binders, illuminators, calligraphers, vellum preparers (a particularly noxious job), and others. Visually speaking, the calligraphers and illuminators had the most immediate and visceral impact upon the reader. It was a slow, labor-intensive process. To speed things up, not every book was illuminated. It is not unusual to find manuscript books that contain only text and no illuminations.

Illustrations could change from one copy to the next. Not all Blemmyae looked the same. Depending on the skill level of the illuminator there could be wild differences even with one text. Sometimes there would be a model book to go by. As a result, it could look like whole armies of monsters existed. Usually, the illuminator simply followed the text and their own imagination. The illuminators, being human like anyone else, might get bored and begin to improvise. Once a main page illustration was complete the illuminator might do doodles in the margins that had little or nothing to do with the text. These doodles could be very intricate and carefully done. They were often animals engaged in various hijinks and often grotesques. Put simply, a grotesque is a drawing that has an unsettling quality to it. Often a human face that is deformed and monstrous (though most scholars argue grotesques can be so different that they defy easy categorization). "The grotesque is not a form or a technique, like stream-of-consciousness. Rather, it is a strategy that attaches itself to different aspects of a literary work – image, character, plot, form, theme."[25]

Startling illuminated works caught readers' eyes. The more intricate the illuminations, the higher the status of the work. Normally when adding illustrations to a manuscript the illuminators stuck close to the text. The illuminators were, after all, human, and sitting hunched over a desk all day drawing devotional pictures the artist would add their own little images, not always authorized or having anything to do with the text. Along with rabbits jousting while riding snails there were knights doing battle with flying creatures. The margins of these works could be densely populated with strange scenes and figures, often monsters. Illuminators loved monsters. They appear often not just as curious little marginal doodles but as magnificent full-page extravaganzas. There are lions with human heads, sea monsters breaching, humanoid primates reading books, and of course the Blemmyae, Sciapodes, and Satyrs. The creatures depicted in these texts were often allegories and stand-ins for other things. Political commentary was less insulting to positions of power if it was done with bunny rabbits or snails. There is also the 14th-century illustration of a monk that bears a startling resemblance to the wise old Yoda of *Star Wars* fame.[26]

With the advent of the Modern period Western collectors had read all the classical authors on monsters. They didn't just want to read about them but wanted to see the strange creatures of myth and legend in the flesh and especially to possess them. They also began to encounter monstrous births of both humans and animals which were real and could be seen. These monstrosities appeared as home-grown creatures. One didn't have to travel to Libya or India to see them firsthand. Coinciding with the spread of printing, stories of monsters came out from the relative exclusivity of illuminated manuscripts in monastery libraries, to written accounts more available to average people. They eventually wound up in museums where anyone could see them. The idea of collecting strange and unusual objects, including biological ones, became a rage with wealthy men with an interest in the natural history of the world, and who wanted to showcase their erudition. They didn't just want to read about exotic creatures; they wanted to hold them and show them off. It is during this period (roughly the 1500s through the 1700s) that the concept of the cabinet of curiosities, or as it was known in Germanic speaking areas, the Wunderkammer was born. This led directly to the modern idea of the library and the museum. Private enthusiasts began to assemble collections for their own study and amusement, and of course as a way to impress their friends. Monsters naturally took top priority, and the stranger the better. A number of categories of objects appeared. There was artificialia, naturalia, and everyone's favorite, exotica.[27] Cabinets even began to appear in the Americas. Thomas Jefferson had one which focused on Native American artifacts.[28]

Beginning with the Protestant Reformation in 1517, Catholic churches came under pressure, even assault. While not Wunderkammers per se, Catholic churches contained relics of saints. These relics came mounted in vessels and boxes (reliquaries) made of gold or silver and encrusted with various jewels as indicators of precious specialness. As the process of the destruction of churches proceeded, rare and valuable objects made it onto the collecting market. Virtuosi exhibited these objects, but to avoid any religious complications, the Catholic theology was stripped away, and the uniqueness, physical value, and rarity were emphasized. These objects slowly made their way into private collections. These acquisitions, however, contained little if anything monstrous.

Then began the slow evolution from the private cabinet of curiosities of the 16th through the 18th centuries created by wealthy collectors to show off their monsters to their equally wealthy friends, to the modern vision of a public museum or library. Some collections were given to local universities or opened to the public with paying admission. They began to take on an educational aspect for more than just the owner and a few elite dinner guests.[29] The collections moved from sensational to educational. The arrangement of objects also evolved. Now, objects in the collection became a more scientific arrangement. Juxtaposing objects to emphasize visual similarities or differences, or just the quirks of the collector, now took on a formal aspect. Teaching became a major aspect of the cabinets. The highlights of these collections, despite their inclination to education, remained monsters and biological oddities. The trend of using monsters as medical teaching aids and the assembling of school sponsored cabinets for the purpose of education continued. Collections of monsters had lost their connections to mythology and theology altogether. They now focused exclusively on explaining the practical aspects of generation, heredity, general medical knowledge, and biology, eventually evolution. Yet the monsters in these collections still held pride of place because of their popularity with museum goers.[30]

One of the early central characters in the cabinet of curiosity world was Ole Worm. A Danish medical doctor, naturalist, and avid collector from Copenhagen, Worm (1588–1654) put together an extensive cabinet.[31] He made major contributions to knowledge of Viking Runes and monuments. He came from family money which allowed him travel across Europe attending various different schools eventually receiving a medical degree in Basel, Switzerland, in 1611. All the time he traveled he collected. Upon his return home he acquired a position as Professor at the University of Copenhagen. He also was named personal physician to King Christian IV of Denmark. This allowed him the time and money to pursue his many scientific interests and build up his collection.[32] He was an early form of what we might call the academic collectors. He had a scientific rationale; he wasn't just collecting "oddities" for the shock value. He installed his collection in his home and produced a catalog finished by his son William as the *Museum Wormianum*.[33] With a special fascination with sea monsters, Worm had a number of dried and stuffed examples of unusual marine creatures.[34] Using his collection he helped show that unicorns did not exist. This legend came about, he argued, from misidentifications of the skeletal remains of the marine Narwhal. His efforts constitute one of the earliest attempts to see the underlining realities of monster myths.[35] His catalog contains many illustrations of the objects from the collection. Plants, flying fish, spiders, shrimp, and of course the Narwhal proliferated in his collection.

Worm encouraged his students to visit his collection and study the contents up close. This, he felt, gave the students firsthand experience handling such objects. He was part of a growing educational movement that argued direct experience with the natural world was preferable to simply reading about it through other people's work in authoritative texts.[36] He arranged his collection in what he felt was a scientifically rational way rather the personally eclectic approach of most of the virtuosi. Committed to practicing medicine as well as teaching, Worm remained in Copenhagen during an outbreak of plague in order to treat the sick in 1664. While doing this he contracted the disease himself and died from it.

Worm was joined in his discussion of unicorns by Jeronimo Lobo (1595–1678) the Jesuit missionary turned naturalist. In *Relazioni Varie* (1693), his description of Ethiopia he recounts his travels to the source of the Blue Nile with the support of the Ethiopian king Susenyos I. He says that unicorns are real and prance around Ethiopia. Lobo ended up losing royal patronage when Susenyos was replaced by a new king who was a follower of the Ethiopian Church.[37]

The Intellectual Virtuosi

In its initial incarnation the fad for collecting strange objects which took hold in the 1500s was in large part more of a social status exercise. Wealthy, elite collectors and art enthusiasts accumulated objects to show off their erudition, sophistication, and wealth as well as for simple entertainment. That idea continued to hold, but now some collectors acquired objects for scientific and educational study.[38] While there are many collectors that could be mentioned here, it can be argued that the greatest collector was the Emperor of the Holy Roman Empire Rudolph II (1552–1612). He had the interest in monsters, but more importantly he had the wherewithal to create an extensive alchemical laboratory as well as a Wunderkammer. In his castle in Prague, he commissioned erotic artwork as well as portraits of deformed humans.[39] Despite the wide scope of his collection and interests, the Emperor was more of a facilitator, hiring alchemists and collectors, rather than a genuine hands-on participant.

Though known primarily as a cosplaying alchemist, Rudolf II was an enthusiastic collector of curiosities and a sufferer from melancholy (depression). He kept these, and his alchemical works, in his castle in Prague. As such he did attempt to collect "monsters" for his Wunderkammer. He put together a *Beastiarium*. Here he kept his taxidermied wonders of odd-looking animals. It wasn't long before he inherited the nearby Schloss Ambras outside

Innsbruck and its extensive collection put together by Archduke Ferdinand II (1529–1595). He added this collection to his own to create one of the greatest Cabinets of Curiosity in the world.

The Schloss Ambras was, and still is, a treasure trove of curiosities. Ferdinand of Tyrol like Rudolf was fascinated by the odd and unusual. The odder the better. His art collection contained portraits of many strange individuals. Likenesses of dwarfs and giants hung next to deformed people as well as the daughter of the "Hairyman": a girl suffering from Hypertrichosis. The collection also contained a portrait of the infamous Vlad Tepes, better known thanks to Irish playwright Bram Stoker, as Count Dracula. With human oddities, deformed persons, it was a bit easier than acquiring a Blemmyae. It was not unusual for the wealthier collectors to hire live dwarfs to act as guides to and even curators of the cabinets. As some collectors saw the cabinets as tools for examining biological questions what better thing than to have an actual live deformed person to look at and ask questions of?

A major draw of the curiosities for collectors was a sense of wonder, for wonder was the keynote of the cabinet of curiosities concept, and the marvels of the collection. The more "wonderful" and monstrous the better. Wonders took the collector, as well as the visitor to the collection, on a temporal journey around the world and back in time. It caused them to ponder where these objects came from and what they meant and eventually led them to questions of relationships and even origins. A number of major collectors, but who were not really simple collectors in the original Wunderkammer sense, began to study these objects from a more modern scientific point of view. While some Cabinet owners collected for the sheer enjoyment of owning strange things. A number of began to publish, not just catalogs of their collections, like Ole Worm, but treatises on biological meaning. Simple ownership took a back seat to study.

Up to this point monster studies were all academic. With a few exceptions they were corralled on the pages of manuscripts only. It was not too long before certain individuals went looking for and collecting actual physical monsters one could see, hold in their hands, and examine. In addition to being subjects of philosophical discussions about the nature of life as well as the structure of cultures, monsters found practical uses in the study of medicine, generation, and the relationship between various animal groups. In addition to philosophizing about the meaning of monsters, a new way of approaching them came into being as researchers and investigators tried to track down and find the physical remains of monstrous creatures. The bodies of monsters, both real and made up, began to appear in museums, medical schools, and in private collections.[40]

John Tradescant

John Tradescant (1570–1638) began his career as a gardener. His patron Robert Cecil, the 1st Earl of Salisbury, sent him on collecting trips in 1610 around Europe for plants he could grow in his extensive gardens.[41] His skills in such a job led him to others as head gardener and garden designer for wealthy people across England. He expanded his wanderings to the Middle East, Algeria, and Arctic Russia. In addition to collecting plants Tradescant's imagination was attracted to non-plant exotica for his own collection. These acquisitions he put into his London townhouse which soon became known to all and sundry as "The Ark." His son, also named John, inspired by his father, became a major collector of natural curiosities as well. Eventually, The Ark had become so full of curiosities of all types that it was opened to the public and not just elite collectors: likely the very first such location in all of Europe, possibly the world. This collection was added to by Tradescant's friend Elias Ashmole (1617–1692).

Producing a catalog of the collection, the *Musaeum Tradejcantianum, or a Collection of Rarities* (1656), Tradescant's interest followed botany though he did dabble in seashells, rocks and minerals, and various oddities. They did not go in for monsters though he did acquire one of the few Dodo birds in the West. This strange creature, only recently discovered by Westerners on the island of Mauritius, ended up with his friend and fellow collector Elias Ashmole. It is currently at the Ashmolean Museum in Oxford. It was in fact Ashmole who published the Tradescant catalog as a sly way of taking the collection away from the Tradescant's wife and adding it permanently to his own. The Ashmolean museum, as it is now known, became the first purpose-built museum open to the public (and still operates as a museum today).

The virtuosi formed communities of discourse to discuss monsters. They contacted one another and even traveled to their cabinets. They exchanged knowledge and on occasion even exhibits. This web of enthusiasts stretched from England to Scandinavia, France and Germany. These collections were considered of such importance by their creators that they left them in their wills so they could be continued. In Italy the apothecary Ferrante Imperato (1525–1615) created a museum out of his collection in Genoa (it is considered one of the very first modern cabinets of curiosity). It was so highly regarded, particularly for its monsters, that scholars came from far and wide to visit it. He also published a catalog of his collection, *Dell'Historia Naturale* (1599). The well-known collector and naturalist Thomas Bartholin (1616–1680), came all the way from Copenhagen in 1644 to stroll amongst its contents. Imperato left the collection to his grandson so that it would be continued. So many cabinets appeared that special works describing the many Wunderkammers were available. Johann Daniel Major's *Vorstellung etlicher Kunstund Naturalien-Kammern in Italien zu Neapolis und Alt-Rom* (1675) took special delight in describing Imperato's collection as a kind of tourist's guidebook.[42]

As Westerners traveled around the world – North Africa and the Middle East, India and by the 16th century the Americas – they gained new views and new knowledge. In a strange way, everything they found was wondrous.

Local people knew these animals and objects intimately, but the new collectors saw new worlds and possibilities. As word spread of Crocodiles and Alligators they became sought after objects. Unusual carved objects and gems and minerals became hot souvenirs for the collectors. Initially, only the wealthiest and best connected could acquire monsters, but eventually less well-off people could as well. Mid-level collectors such as doctors, university professors, and naturalists gained access to the curiosity stream.

The Crocodile, and eventually the Alligator, became important totems of position to the virtuosi. They represented monsters more readily available for acquisition than centaurs or sciapods. By having one taxidermied and hung from the cabinet's ceiling it took pride of place in the collection. Often the first thing a visitor would notice, the exotic crocodile seemed a visceral connection to the outside exotic world. These creatures became many things to those who displayed them. In its simplest form, curiosity collectors and virtuosi acquired crocodiles as status symbols. The whole point for many virtuosi and their Wunderkammers was to show off their erudition and knowledge of the natural history of the world. What better way to do that than to boast of such a rare – at least in Europe – beast. Not only popular with virtuosi, crocodiles came to inhabit the spaces of alchemists and churches as well.[43] As Spencer J. Weinreich points out, "The animal was not merely a symbol," but a channel through which to transmit or a receptacle in which to store ideas about the very nature of life on Earth.[44] The creature took on many meanings for different audiences. Its very monstrosity suggesting questions of its origins, its relationship to other animals, and its other worldliness. When the Americas became known to Europeans many strange animals and plants came to their notice. Alligators in particular drew attention. So much like crocodiles yet different, the Alligator created new questions about the relationship between Old and New Worlds. Did the differences mean a different creation? The discovery of the New World, not mentioned in the Bible, raised many questions about why it wasn't. Did the Alligator, and even the Crocodile for that matter, undermine knowledge about the organization of life on Earth or did it support the Biblical worldview? Did these monsters bring the Old and New Worlds closer together, or push them further apart?

Reproducible Monsters

With the advent of printing, inexpensive pamphlets and single sheet broadsides began to appear. As with manuscripts, monsters immediately made very popular subjects. In 1646 one such *A Declaration of a Strange and Wonderfull Monster* made the rounds in England. The star of the show was the monster prominently displayed in a woodcut. The creature was the result of a human birth. It made a connection to the past as it looked very much like a Blemmyae. A few years later in 1652 another pamphlet told of a deformed pig. The text explained the reason this hideous creature was allowed to be born was to prove to people "we would not be frightened by His Almighty power." Monsters were messages from God reminding, mostly poor, people to stay in line and not question the church, their pastors or their betters.[45] Deformed pigs were popular. A two bodied pig with a single head came out in a German text which was so popular that an English-language version was released as well.[46] The ability to reproduce line drawings meant illustrations of monsters could appear in cheaply made publications relatively inexpensively. This brought monsters to a wider reading public which developed a taste for the monstrous that remains to this day and shows no sign of abating. The tabloids of the 21st century are a direct throwback to the sensational illustrations of the 17th century.

The Usual Suspects

The monsters of India and Persia seeped into the Western consciousness via authors such as Herodotus, Ctesias, Megasthenes, and Pliny. Creatures such as the Blemmyae, Sciapodes, Cynocephaly and others became the usual suspects all later authors on monsters employed. From that point on, every illuminated manuscript as well as printed works included renditions of these monstrous races. Having established, at least in their minds, that monsters and monstrosity were important for balance in the universe and to understand the great diversity of life on Earth, naturalists interest in monsters grew just as rapidly as it did for the general public.

And around and around we go, just like yesterday.

Notes

1 Filippo Del Lucchese, *Monstrosity and Philosophy: Radical Otherness in Greek and Latin Culture* (Edinburgh Scholarship Online, 2019), 93–129.
2 Lucchese, *Monstrosity and Philosophy*, 1–7.
3 Andrea Falcon and David Lefebvre, eds., *Aristotle, the Generation of Animals: A Critical Guide* (Cambridge University Press, 2018), 207 and 427.
4 Phillipo Del Lucchese, *Monstrosity and Philosophy: Radical Otherness in Greek and Latin Culture* (Edinburgh University Press, 2019).
5 Lloyd Llewellyn-Jones, *Ctesias' 'History of Persia': Tales of the Orient* (Routledge, 2012).
6 The Blemmyae were known by several different names and variations of spelling.
7 Thorfinn Karlsefni, one of the reputed Norse discoverers of North America claimed he saw a Monopod in Vinland in around 1010 CE. Gisli Sigurdsson (Introduction), *The Vinland Sagas* (Penguin Classics, 2008).
8 Megasthenes, *Indica* (Routledge, 2023). For a more in-depth discussion of the role of Greek interpretations of life in India see, Rudolf Wittkower, "Marvels of the East: A Study in the History of Monsters," *Journal of the Warburg and Courtauld Institutes* 5 (1942): 159–97.

9 Augustine of Hippo, Henry Bettenson, trans., *City of God* (Penguin Classics, 2004), Book VIII.
10 Ovid, *Metamorphoses*, trans. Bernard Knox (W.W. Norton, 2004).
11 See, Debbie Felton, ed., *The Oxford Handbook of Monsters in Classical Myth* (Oxford University Press, 2024), 230–42.
12 Paul Cartledge, ed., trans., *Herodotus: The Histories* (Penguin Classics, 2014).
13 James Redfield, "Herodotus the Tourist," *Classical Philology* 80, no. 2 (April 1985): 97–118.
14 *Histories*, book IV.
15 *Pliny's Natural History in Thirty Seven Books* (Lagare Street Press, 2022).
16 Paula Findlen, *Possessing Nature: Museums, Collecting, and Scientific Culture in Early Modern Italy* (University of California Press, 1994), 3.
17 For an excellent discussion of the Monstrous Races in European thought see; John Block Friedman, *The Monstrous Races in Medieval Art and Thought* (Syracuse University Press, 2000), and David Williams, *Deformed Discourse: The Function of the Monster in Medieval Thought and Literature* (McGill-Queen's University Press, 1996).
18 Lucretius, *On the Nature of Things*, trans. Martin Ferguson Smith (Hackett, 2001).
19 Lucretius, *On the Nature of Things*, 120.
20 Lucretius, *On the Nature of Things*, 8.
21 Special thanks to Dr. Xurong Kong for help with the history of monsters in China. Xurong Kong, *Fu Poetry Along the Silk Roads: Third Century Chinese Writings on Exotica* (ARC Humanities Press, 2021).
22 Kenneth J. De Woskin and James Irving Crump, trans., *In Search of the Supernatural* (Stanford University Press, 1996).
23 Anne Birrell, trans., *Classic of Mountain and Sea* (Penguin Classics, 2000). Written sometime in the 4th century BCE, by an unknown author, during the Han Dynasty (202 BCE- 220 CE).
24 References here are all taken from John Ashton, trans., *The Voiage and Travayle of Sir John Maundeville Knight* (London: Pickering and Chatto, 1887) and C. W. R. D. Moseley, trans., *The Travels of Sir John Mandeville* (London: Penguin Books, 1983).
25 John C. Van Rys, "Reclaiming Marginalia: The Grotesque and the Christian Reader," *Christianity and Literature* 44, no. 3/4 (Spring–Summer 1995): 345–57.
26 Karl Shuker, "From Star Wars to Smithfield, the Curious Case of the Medieval Yoda Monk," *Shuker Nature*, December 17, 2015. On-line Blog. Also see, Damien Kempf and Maria Gilbert, *Medieval Monsters* (British Library, 2015).
27 Caroline Galambosova, "What Is a Wunderkammer?" *Daily Art Magazine*, September 23, 2024.
28 Joyce Henry Robinson, "An American Cabinet of Curiosities: Thomas Jeffersons Indian Hall at Monticello," *Winterthur Portfolio* 30, no. 1 (Spring 1995): 41–5.
29 Patrick Mauries, *Cabinets of Curiosities* (Thams and Hudson, 2002).
30 Sabrina Richards, "The World in a Cabinet, 1600s," *The Scientist* 26, no. 4 (April 1, 2012): 88.
31 Early 'cabinets' were boxes rather than rooms: elaborate arks into which the collector placed his prized possessions. The 'cabinets' were often elaborately constructed with fine inlay and draw pulls of rare materials. It was in the 16th century that these cabinets became full sized rooms and even entire buildings.
32 Rafael Romero-Reverón and Luis A. Arráez-Aybar, "Ole Worm (1588–1654) – Anatomist and Antiquarian," *European Jornal of Anatomy* 19, no. 3 (2015): 299–301.

33 Ole Worm, *Museum Wormianum: Historia Rerum Rariorum* (Lugduni Batavorum, 1655). As many serious collectors did, Worm published a catalog of his collection which included one of the most famous and reproduced illustrations of the interior of a cabinet we have.
34 Maggie Boyd, "From the Archive," *Bulletin of the American Academy of Arts and Sciences* 76, no. 2 (Winter 2023): 68.
35 In 1927 the American horror writer H. P. Lovecraft, in his *A History of the Necronomicon*, claimed fictionally that the first Latin translation of this legendary book of Black Magic was written by Olaus Wormius.
36 Fiona Melhuish, *Ole Worm's Museum Wormianum* (University of Reading Special Collections, May 12, 2020).
37 Jeronimo Lobo, *Relazioni Varie* (Florence, 1693).
38 MacGregor Impey and Arthur Oliver, *The Origins of Museums: The Cabinet of Curiosities in Sixteenth-and Seventeenth-Century Europe* (Oxford: Oxford University Press, 2001).
39 R. J. W. Evans, *Rudolf II and His World: A Case Study in Intellectual History 1576–1612* (Thames & Hudson, 1997).
40 Peter Dendle and Asa Simon Mittman, *The Ashgate Research Companion to Monsters and the Monstrous* (Routledge, 2012).
41 Prudence Leith Ross, *The John Tradescants: Gardeners to the Rose and Lily Queen* (Peter Owen Publishers, 2006).
42 James Clifton, "Kunst- und Naturalien-Kammern of Seventeenth-Century Naples: Johann Daniel Major's view from Germany," in *The Science of Naples* (UCL Press, 2024).
43 Spencer J. Weinreich, "Thinking with Crocodiles: An Iconic Animal at the Intersection of Early-Modern Religion and Natural Philosophy," *Early Science and Medicine* 20–3 (2015): 209–40. (No self-respecting alchemist was without a crocodile or other similar monster hanging from their ceiling).
44 Weinreich, "Thinking with Crocodiles," 210.
45 Katherine Park and Lorraine Daston, "Unnatural Conceptions: The Study of Monsters in Sixteenth and Seventeenth Century France and England," *Past & Present* 92 (August 1981): 20–54. Also, Serina Patterson, "Reading the Medieval in Early Modern Monster Culture," *Studies in Philology* 111, no. 2 (Spring 2014): 282–311.
46 Sheila O'Connell and David Paisey, "This Horryble Monster: An Anglo-German Broadside of 1531," *Print Quarterly* 16, no. 1 (March 1999): 57–63.

2

COLLECTING MONSTERS

There is considerable modern scholarly debate concerning the transition from the medieval to modern eras, the advent of the Scientific Revolution and whether these convenient terms even apply.[1] A detailed examination of the complex combination of events and ideas responsible for this change is beyond the scope of this study. However, interest in animal studies ensued during the 16th century that brought scholars back to the realm of monsters as genuine biological entities rather than simply models for teaching proper Christian behavior. While not supplanting the theological and metaphorical use of animals that dominated the medieval period, the biological worthiness of living things as topics of study became an option picked up by a number of authors.

Several trends helped make monster studies a pursuit for a cadre of naturalists. Manuscript copies of the works of classical authors became more available in Latin translation. Such texts were already known but began to be appreciated on a more prosaic level by intellectuals interested in the nature of things. Printing appeared and became more efficient and paper making improved, allowing for the production of many more volumes and wider circulation. Print technology allowed not just for the production of scholarly books, but the cheap pamphlets and broadsides that appealed to a more popular audience. Religious tracts used monsters to scare the faithful into proper behavior, and to direct animosity and distrust toward those denominations' leaders deemed problematic. Local stories of monstrous births became the grist for religious and political hate speech that poured out as cheap tracts of often a few pages or even just a broadside became widespread. This period also saw the Protestant Reformation and the discovery of the Americas. With new technologies of communication, new information came in from new

DOI: 10.4324/9781003488767-3

lands. The study of plants, animals, fishes, and monsters made great strides.[2] Monsters also made it into the domain of the learned to be pondered in the erudite realm of the private study, and by the masses in the rough and tumble of the street with monster broadsides glued to walls in public places.

Along with works by classical authors, a work called *Physiologus* served as a well-worn source for authors of the monstrous. A manuscript by an unknown author, with an uncertain date of original production, the *Physiologus* (The Naturalist) stands as an early compilation of animal lore. It first appeared in Greek and by 700 AD in Latin, then eventually in other languages. It was widely known, read, and influential to authors and artists alike throughout the medieval and then modern period. Along with descriptions of various animals – including several mythical beasts like the Unicorn – the text contained disquisitions on the metaphysical meaning of animals in the time and tested style of medieval manuscripts.[3] The foundation knowledge contained in the book originated in Indian, Egyptian, and Hebrew as well as Greek and Roman lore. The compiler of the text took early stories about animals and ran them through a Christian sensibility. Where monster stories had begun to turn from theological to biological, this work tried to steer monsters back to the theological. The author turned the purpose of the texts away from the natural history of the ancients into a commentary on Christian metaphysics with animals standing in as human subjects.[4]

The behavior of the animals in the *Physiologus* stood as models for human behavior, not unlike the tales of Aesop. They showed strength, weakness, repentance and other supposed Christian virtues and shortcomings. They also showed anti-Semitism. For example, the Owl is "the figure of the Jewish people who, when our Lord and Savior came to save them, rejected Him." The phoenix flew as the symbol of the resurrection of Jesus, but "the Jews were angered by his words" about the resurrection. "How foolish are those men (the Jews)," the author says, "who grow angry at the words of the Savior."[5] Some early Modern authors took the model of the *Physiologus* text and turned the subject matter yet again back into natural history. However, the metaphysical use of animals and monsters never did quite disappear. Naturalist monster authors had to compete with a religiously based and sensational monster publication trend that far from dissipating became even more potent and widespread.

The Monk Calf and the Popular Monsters

While monsters populated medieval manuscripts in profusion, they also leap to the pages of printed matter. Wood block and movable type printing was established in Europe by the 1450s and by the early part of the next century monsters were being printed as well. They would get their widest acknowledgement through the spread of pamphlets and tracts which could be produced quickly, cheaply, and in numbers. One of the more popular monsters depicted during this period, the Monk Calf, first appeared in a pamphlet by Martin Luther and Philip Melanchthon. *The Papal Ass of Rome and the Monk Calf of Freyberg* (1523) included an illustration of the purported creature, a pig-like monstrous birth on two feet with layers of flesh hanging off it. An earlier equally popular version of such a creature which may have inspired Luther, the Ravenna Monster, also had metaphysical rather than biological implications. Based upon a grossly deformed infant from the Ravenna area, born sometime between 1510 and 1512, said to indicate God's displeasure with the Italian people, it appeared in several different publications including Stephen Batman's *The Doome, Warning All Men to the Judgments* (1581). The Monk Calf provided no subject of biological study, but rather a vehicle for religious and political commentary, and persecution. It fitted the anti-Catholic diatribes and prurient rumor mongering of the lower orders meant to win converts to the Protestant cause by turning them away from the establishment. The skin of the Monk Calf was depicted as looking like the cowl and religious garb of a Catholic monk. Luther used the creature as a metaphor of the Church he hated so much and the displeasure the Church caused God. "Monsters," Luther and Melanchthon said, "most times doe note and demonstrate unto us the ire and wrath of God."[6] Monk's were accused by anti-Catholic zealots like Luther (who himself had been both a Catholic and a monk) of trying to achieve Godliness through the donning special clothing instead of genuine spirituality. Monk's lives, Luther claimed, "consist of nothing but gobbling food, of drinking, and of sex".[7] Thus, they were not true religionists, but ghastly imposters just like this malformed monstrosity. The Monk Calf was, Luther said, God's way of producing a creature to show all and sundry the perfidy of the Church.[8]

Luther's pamphlet on the Monk Calf proved a popular model for later authors who used monsters as hammers against the Catholic Church and to remind readers of God's ever-present power. The popular monsters, those designed for consumption by the masses, came in several categories. The more rabid tracts, like Luther's, attacked the Catholic Church rather than discussing strange births. These authors had no interest in natural history or human generation. The Church and its functionaries such as priests, bishops and naturally the Pope were the monsters. The creatures depicted in these pamphlets could sometimes be outrageously made-up concoctions along with the usual two-headed babies or deformed pigs. Luther's Monk Calf was only the most widely known and copied. Such tracts included *Novembris Monstrum* (1641) and *A Vindication of the Truth of the Christian Religion* (1694) in which variations on the Monk Calf appeared.

The other most common category of monster tract included those which used monsters as a way of showing the wrath that God would bring down, not only on Kings and countries but also on commoners as well. These can be collectively known as scare tracts. The author of *A Declaration of a Strange and Wonderful Monster* (1645) saw this particular creature as "a wonderful manifestation of God's anger against wicked and profane people." The Protestant mother in the story had given birth to a monstrous child, the author claimed, because God had damned her for marrying a Papist (Catholic) man. Her poor choice of husbands brought on this terrible calamity. Other Protestant women should, therefore, be careful of who they pick as the father of their children.

A pamphlet of 1652 on the birth of a monstrous pig said, "these strange and monstrous things which Almighty God sendeth amongst us that we would not be frightened of his Almighty power." In another a child born in Verona supposedly grew from a baby so rapidly it achieved the size of an adult in just a few months. This "child" then began to spout prophesies about the downfall of various countries including France and Poland. "The Inquisition," the author states, "has since taken the child into their custody." Non-human monsters appeared in these pamphlets as well. In pamphlets such as *A Certaine Relation of a Hog-Faced Woman* (1640), *Great News from Saxony or a New and Strange Relation of the Mighty Giant Koorbmep* (1680), and *A Most Strange but True Account of Very Large Sea-Monster* (1704) theologians and other self-appointed moralists felt obliged to remind their audiences – of mostly poor and barely literate peasants and working-class people – that God watched them in their little affairs and would bring down swift and terrible judgment on them in the form of a monster if they did not obey their Protestant Christian betters. Only rarely did the monster pamphleteers single out a noble or wealthy family. It was the lower classes who were the most prone to such a calamity.

The wake of the Protestant Reformation saw an increase in not only popular, sensational pamphlets in the style of Luther's but also in more scholarly tomes. Some of these, including Conrad Lycosthenes' *Prodigiorum ac Ostentorum Chronicum* (1557), and Pierre Boaistuau's *Histoires Prodigieuses* (1559) continued the trend of monsters as omens of things to come or of the perfidy of the Catholic Church being pointed out by God, but also leaned more in the biological direction. Less political and more natural history oriented were Guillaume Rondelet's *Libri de Piscibus Merinis* (1554), and then Ambroise Paré's *Des Monstres* (1573). Whether meant to skewer the church or Jews or to elucidate the reality of wondrous creatures, these books brought monster studies to a wider audience than ever. As time went by, however, the view that monsters represented portents of evil times or God's wrath, like a comet or earthquake, went out of use in all but the most popular and sensationalist forms of books and broadsides. Renaissance era Humanist naturalist monster enthusiasts and collectors of curiosities of various stripes in Europe and Tudor England still relied upon Grecco-Roman sources, combining with it that sense of Christian morality, but increasingly viewed animals and monsters as topics of biological wonder as well.[9]

Two groups took scientific interest in monsters: Physicians and naturalist collectors saw monsters as vehicles to understand the natural world. The general notion included the idea that abnormalities and unusual body shapes could be used to study things like generation and the general order of nature. While individual members of these two groups still held a theological component to the meaning of monsters, they worked to see a more naturalistic and scientific view of these entities as well. This interest spurred them on to examine the biological underpinning of life. As monsters were so different it was thought their differences could shed light on the process of life. Through this interest in the mechanics of procreation and heredity, biological diversity and systemization would be better understood as would growing questions of transmutation. The 17th century also saw the spread of the experimental philosophy and empiricism (borrowed from Islamic alchemy), with its emphasis on direct observation and study: an approach that would simultaneously bring monster studies to new heights of rigor, but ultimately undermine it.[10]

The Cabinets Are Opened

Interest in collecting the natural world remained a mania amongst wealthy educated elites in 16th- and 17th-century England and Europe. Well-to-do scholars, physicians, apothecaries, and dilettante nobles with the money, time, and sophistication to appreciate it, created private museums and collections to house and showcase examples of wonders and monsters. In Italy the place where an erudite man kept his collection and studied and pondered it came to be known as *Studioli*, while in Germany by the 1550s *Kunstkammer* eventually *Wunderkammer*. European collectors came relatively late to the concept of storing books (in their many guises) on shelves in specially constructed rooms or even buildings. Such practices had already been going on for several centuries in China, Korea, and the Islamic world.[11]

Some of the scholarly monster enthusiasts helped found the modern sciences of biology, botany and zoology. Men such as Conrad Gesner, Ullisse Aldrovandi, Fortunio Liceti, and others sought to concentrate all the world's living things, especially unusual things, in the confines of their private libraries or between the covers of a book: to wrestle under control life itself.[12]

Not all the cabinets were the same. Some reached heights of great opulence and extent with the collection filling entire floors of castles, while others filled only a single custom-built wardrobe-like box. Manfredo Settala of Milan built a collection so large it took seven volumes to catalog it all. Naturalists of modest means assembled smaller collections while large scale Wunderkammers came together through the collecting impulse and wealth of flamboyant nobles. A network grew up as collectors sought to show off their collections and see the pieces acquired by fellow virtuosi.[13] There were two types of these collectors. Scholars who wanted collections as study aids and pure Virtuosi more interested in monsters as prizes to be shown off to friends and nobles rather than study objects. It was a way to establish social status. Just as there were hierarchies of collectors there were hierarchies of specimens to be collected. Certain objects counted more than others. Monsters were at the top, especially human monsters. Live specimens were the pinnacle, though skeletal parts were next. In a pinch a painting of a human monster would suffice. Dried skins, skeletal parts, and body parts pickled in jars of a monstrous creature rated very highly. One sure sign of exclusivity was the stuffed crocodile suspended from the ceiling of the studio (later this was knocked down a peg when stuffed alligators began to arrive from the Americas). Alchemists too engaged in the ceiling mounted monster motif. And while elephant ivory was highly prized for its own or as the basis for highly complex puzzle sculptures, it paled compared to the mounted narwal tusk.

Collections of sea shells, skulls of cats and dogs lined up next to stuffed sea monsters, pickled two-headed babies, and fossil fish, the more exotic and wondrous the better. Anomalies like monstrous births of both humans and animals, exotic crystals or rocks that looked like human faces, and artwork depicting such things filled these collections. "The museum," as historian Paula Findlen says, "was designed to represent nature as a continuum."[14] All life, the scholars argued, fit into a scheme, a great scale that showed God's handiwork, with all living things in their proper places. The continuum had spaces in it that needed to be filled. Monsters could both fill those gaps, but also create them, posing difficult questions not easily answered. These collectors and their close cousins the encyclopedists worked to make established animals better known and to bring others out of the realm of the fantastic, while eliminating those of pure fable.[15] The wealthier and more grandiose collectors assembled Wunderkammers for the fascination with strangeness and the collectability of such items, not unlike the way wealthy collectors today bring together vintage cars or paintings: to show off. However, a number of collectors did so as a way to scrutinize the nature of the human body by having a study collection.

The Monster Doctors

Naturalists comprised the largest group of monster enthusiasts during the Renaissance. However, physicians and surgeons also took notice of how monsters, particularly monstrous births, influenced human generation and reproduction. While a number of texts appeared from medical men who included discussions of monstrous births – especially works on gynecology and midwifery – two stand out during this period. Jacob Rueff (1500–1558) worked as city physician at Zurich and instructed midwives, though he considered women lesser versions of men. His *De Coneptu et Generatione Hominis* (1554), aimed at physicians, surgeons and particularly midwives, covered the techniques of childbirth. As with so many authors of early books on obstetrics and gynecology, Rueff saw monstrous births as fascinating subjects that could bring insight into generation and heredity and were an easy way to bring non-specialist readers to the text.[16] (Books on child birth, such as the wildly popular *Aristotle's Masterpiece* which still saw print into the 19th century, continued to carry illustrations of monstrous births when virtually nothing else from the text had accompanying pictures). Rueff based his book in part upon Andreas Vesalius' *De Humani Corporis Fabrica* (1543) and Conrad Gesner's *Historiae Animalium* (1551–1558), as well as an earlier obstetrics book *Der Swengeren Fruwen und Hebammen Rosengarten* (1508). He subscribed to the divine origin of monsters as when he described one particular deformity as having originated from God because of His displeasure with sodomy.[17]

The other important author dealing with obstetrics and monsters was the Frenchman Ambroise Paré. While the story of his early life is a bit muddled, Paré (1510–1590) began his career as an apprentice barber-surgeon. He then joined the French army as a field surgeon and saw extensive battlefield service in a number of the many religious wars which ravaged Europe during the Reformation. Amidst the blood and death of combat he saw the futility of the traditional employment of hot oil and cauterization for sword cuts, gunshot wounds, and amputations. He developed an alternative method which abandoned cauterization in favor of surgically tying off arteries to stop blood flow. His method proved exceptionally successful and the survival rate of wounded soldiers under his care improved. He rose steadily up the ranks of French medicine by making a number of choice associations and friendships.[18] He eventually became chief surgeon to Henri III and counted the political philosopher Montaigne a friend.[19]

Along with surgical techniques, Paré pioneered the design and implementation of mechanical prosthetics for amputation patients. Paré wrote on obstetrics and gynecology and helped develop techniques for inducing labor without drugs. He believed birth defects such as club foot resulted from the mother sitting in an odd position, or being injured, in which the fetus receives an injury in utero. He seems a paragon of the increasingly influential new secular science and experimental philosophy. However, Paré is a fascinating example of the border regions between theology and science which is a hallmark of this age. While his actual religious affiliations and sympathy – between Catholicism and Protestantism – are debated by scholars today, he lived and wrote as a deeply religious man. He famously said that he treated his patients, but God cured them. He felt doctors could only cure natural disease ailments not supernatural ones caused by sorcerers or demons.[20]

Like Rueff, Paré wanted average people, even women, to know of medical procedures and the techniques of birthing. This attitude did not endear him to the Paris medical establishment. His monster work – which at the time was thought of by some as an unimportant and needless distraction from proper medical issues – also did not impress his colleagues. Historian A.W. Bates argues that Paré included this material in his text as a way of boosting sales of his book outside the medical community.[21]

Paré's interest in monsters stemmed from his curiosity about human generation, but also a general interest with the outré. He once claimed to have encountered a Toad-in-the-hole. An unverified phenomenon – which still has proponents today – Toad-in-the-Hole events are when seemingly solid stones are broken open to reveal a fully formed toad or frog alive inside. Paré's encounter with just such a specimen began him thinking about spontaneous generation and how monster studies might bring some insight into how living things could appear without any apparent procreation being involved.[22]

Paré began working on his magnum monster opus, *Des Monstres et Prodiges* in 1570 as a way of gaining a better understanding of human reproduction. He wanted to understand what caused the abnormal instead of the normal. "Monsters," he said, "are things that appear outside the course of nature" while "wonders" went against nature.[23] He argued that God allowed for the deviation yet there were natural reasons why deformities came about. He was unwilling to simply fob off monstrosity on God completely. He began the book by articulating thirteen reasons monsters appeared. First and foremost, monsters indicated God's glory or his wrath. This, however, accounted for only a few instances. The majority of monsters, such as babies with more than one head or sets of limbs, appeared as a result of either too much or too little "seed." They could be caused by some physical problem with the mother's reproductive system, or some injury to the mother during pregnancy which might cause a deformed skull. He added finally that they could be caused by the machinations of "demons and devils" and "wicked spital beggars." Despite his references to the supernatural, Paré worked to establish a more naturalistic set of understandings and explanations. He added representations of monsters including the Monk Calf as ways of explaining their origins in biological terms. Paré extended his discussion of monstrous creatures to his work on surgery as well. In *Opera Chirurgica* (1594), along with drawings and discussions of mechanical hands and legs, Paré presented the image of a griffin-like creature, the Succanth.

Increasingly, authors like Paré began to reject the religious and political trappings of monster studies and leaned to scientific and prosaic explanations and examinations, but could not quite let go of all the old ways.[24] Guillaume Rondelet's *Libri de Piscabus Marinis* (1554–1555) included his direct observations of fish and other aquatic life, yet too made reference to Luther's Monk Calf. Despite his apparent "scientific" approach, Paré wrote in the popular tradition of wonders and religious pamphleteers. In describing a monster that came out of a chicken egg in the home of a barrister in March of 1569, Paré said, "The present monster you see depicted here was found inside an egg."

Like some erudite carnival barker Paré goes on to say that it had "the face and visage of a man, his hair made up of little snakes." Then for real impact he finished by exclaiming that this abomination was "completely alive."[25] As learned authors began to lean toward a less theological stance on monsters; they still clung – unconsciously or not – to the superstitions of their day.

The monster writers of the 16th century, whether physician or naturalist (many were both), followed a pattern of organization for their books. They described the animal then told stories about them often taken from mythology and ancient texts. At least one writer did something a bit different. Frenchman Pierre Belon (1517–1564) eschewed the tradition of natural history writing by using drawings of animal skeletons to illustrate his work in addition to the drawings of the creatures alive. In the 1550s he took a comparative anatomy approach: something all but unknown to his contemporary virtuosi and collectors. His books on fishes, birds, and mammals pointed out the similarities between species. He included a drawing of a human and bird skeleton positioned side by side. Like Jacob Rueff he took inspiration from Vesalius' influential book on human anatomy *De Humani Corporus Fabrica* (1543) which featured exacting renderings of human anatomy.[26] He worked to turn the approach Vesalius had employed with human anatomy to animal structure. While Belon stood apart from his contemporaries, a few authors held an especially influential place in monster studies.

Conrad Gesner

Of the zoological and monsterological authors of the 16th century, Conrad Gesner stands out. A Swiss naturalist and botanist Gesner (1516–1565) studied at Strassburg and Bourges. By 1541 he earned a medical degree at Basil, after which he settled down in Zurich and did extensive work on the systemization of plants. As a close friend of the Englishman and Court Physician, John Caius of Cambridge, Gesner went into the field to study plants directly, and even on occasion questioned Aristotle.[27] He was an avid encyclopedist and bibliographer who compiled important lists of scholarly books in the humanist's languages of choice, Hebrew, Greek, and Latin.[28] Gesner wrote a number of books on various forms of animal and plant life, but his great masterpiece is the five volume *Historiae Animalium* (1551–1558). In the tradition of the virtuosi and encyclopedists, Gesner included a wide range of animals both known and unknown like the rhinoceros and the unicorn. He included details both confirmed and mythological.

Despite the magisterial size of the *Historiae Animalium* and the widespread influence it would come to have on monster studies, the majority of the work is taken up by discussions and descriptions of known, non-monstrous creatures. Only 10% of the project concerns monsters of any kind. Each volume of the set covered a different topic. *De Quadrupedibus* covers known animals such as cats, dogs, camels, cows, and elephants, for example. He does include the unicorn along with a [Sumatran] Rhinoceros and refers often to Pliny and Herodotus. In *De Piscibus* Gesner includes an illustration of a giant lobster attacking a man along with a fantastic rendering of a Narwhal. In *Historia de Serpentibus* he describes the Basilisk and dragons. Here he gets deepest into monster lore and the realm of dragons. One set of drawings has a dragon that is essentially a snake but with jowls. The next adds leather wings while the third includes clawed reptilian legs, two horns, and an arrowhead shaped tongue. This body style, while not original to Gesner, had a long shelf life as the most copied species of dragon in Western texts. The image of the creature Gesner produced closely resembled a dragon drawn in a Medieval French bestiary from 1270, right down to the wagging arrow-headed tongue.[29] This model dragon appeared again and again in later books by many authors. Another illustration has the Hydra with seven heads, perched atop each is a royal crown. Then a huge marine dragon pulls a ship under. Gesner refers to the Norwegian author Olaus Magnus as well as the medieval Moslem physician-alchemist Avicenna on the toxicity and medical uses of dragon parts.

The largest concentration of monster-related lore by Gesner comes in *Aquatilia*. In this volume another giant lobster attacks a man. A series of drawings of whales – not always accurately depicted – along with rare early depictions of the whaling industry are included. These are followed by monstrous squid and octopi. Since no self-respecting author on monsters do without it, Gesner includes the ever-popular Monk Calf, and for good measure the crowned, seven-headed Hydra reappears.

Gesner's work was widely influential to other naturalists. Indeed, later authors would simply copy out entire sections of his work for their own. Like Pliny before him, Gesner was lauded for the breadth of his knowledge. He was even referred to as the New Pliny. Not all his readers were quite so impressed. As a Protestant and Godson of the influential Reformation leader Ulrich Zwingli, Gesner's work was condemned by the Catholic Church who placed the *Historiae Animalium* on its list of forbidden books. Gesner himself died of the plague. His work lived on long after his demise, however.

Despite his dubious religious relations, Gesner was used as a source work for theological monster exploiters. A religious tract utilizing Gesner's – and other naturalist's work – by Stephan Batman appeared in 1581. *The Doome warning all men to the iudgemente wherein are contayned for the most parte all the straunge prodigies hapned in the worlde, with diuers . . . conuersion towardes God* used the creatures from Gesner but stripped them of their natural history in favor of sermonizing over their prodigious qualities. The monsters – whether human, animal or celestial – represented were real and had terrible consequences for man. However, he seems to argue that an easy way to avoid them is to simply ignore them. Paying attention to monsters, Batman warned, was the wrong approach. Subtly ascribing the power of portents to the believer Batman suggests that prosperity and Godliness for both individuals and nations will result from rejecting portents just as embracing them will bring on catastrophe. His own careful studies of monsters shows that "so manye prodigious births, starres of unaccustomed appearance . . . from which proceed pestilence, plague, war [and] hunger" are many[30] He made his "whole paineful studie" only because God allowed him to without retribution. The retribution for sins against God was still a major theme in the discussion of monsters and portents. Batman includes many examples of the role of monsters in how bad people and tyrants get their just deserts. One story Batman uses as an example concerns a young woman known locally for her promiscuity. While riding her horse lighting struck and killed her. Inspection of her body revealed her tongue "was standing out of her lower partes."[31] This bizarre post-mortem condition was, Batman explained, a sign of what happens to women who do not remain chase. He also adds discussions of the Monk Calf and Ravenna monster.

Four-Footed Beasts

Gesner became a towering zoological authority copied by many later writers, often as blindly as Aristotle and Pliny had been copied before them. An example of the border region between the collectors and the growing reliance upon direct observation is Edward Topsell's *Historie of Foure-Footed Beastes* (1607) and his follow-up *Historie of Serpents* (1609).[32] A theologian rather than a naturalist Topsell (1572–1625) wrote on religious issues, but turned his attention to the world of animals as a way of showing the hand of God in the diversity of living things.

Writing in English rather than Latin, Topsell mostly copied out Gesner's *Historiae Animalium*, and added some new material about creatures discovered more recently, particularly in the Americas. He still included a good bit of mythological and fantastic creatures relying upon Biblical passages to explain the behavior of some species, yet he occasionally questions legendary accounts and descriptions. Added to this were illustrations, from several different sources, including a very man-like, tailed Satyr playing a flute. Mostly, he took past authorities at face value mentioning how apes, for example, were terrified of snails. In the *Historie of Serpents* he claims that "the dragons of *Phrygia* when they are hungry, turne themselves toward the west, & gaping wide, with the force of their breath doe draw the birdes that flie ouer their heads into their throats."[33]

Topsell's opening pages are turned over to anthropomorphizing the lore of apes and munkys [his spelling]. In explaining the different names given to primates, like Cereopes, he says such terms came about "because of the wicked crafts, deceits, impostures and flatteries" these creatures engage in, yet at the same time points out that the Cynocepali were in fact "babouns." Arguing against the common misconception that baboons were a form of human being he says "it is the error of vulgar people, to think that babouns [sic] are men, differing only in the face or visage."[34]

Topsell complained that his work was not widely read or appreciated. He told his patron, Lord Ellesmere, that the book did well financially for the printer, but had left him impoverished.[35] It was used, however, to inspire artisans as well as naturalists. In the early 1660s the Devonshire plasterer John Abbott II (1639–1727) produced a pattern book for fellow craftsmen. These books served as guides to artists in creating interior decorations for houses. It included illustrations of animals and monsters taken from Topsell's *Foure-Fotted Beastes*.[36] Baroque artists, architects and sculptors made extensive use of books like Topsell's to create the heavily leaden and dramatic iconography this art movement was known for. Baroque practitioners – mostly Italian, and mostly uninterested in monsters as living things – turned monsters into frightening creatures and demons to adorn paintings and building facades.

As a theologian rather than a naturalist Edward Topsell could only copy Conrad Gesner and then slip in new material from other authors with a bit of Christian worldview sprinkled in. He made no original contribution to the literature. The other titan of Early Modern monster studies did some copying and paraphrasing but went further.

Ulisse Aldrovandi

Italy was one of the primary centers for the development of natural history studies. Along with Gesner the other important and influential Renaissance collector with an interest in monsters was the Bolognese naturalist Ulisse Aldrovandi (1522–1605). Slated for a business career by his noble family Aldrovandi became enthralled by natural history while a student at Bologna and then Padua and while in prison as an accused heretic over Trinitarian beliefs. He went to medical school and eventually became professor of philosophy at Bologna in 1561. His interests ranged widely and he is thought to be the first to use the term geology.[37] He also did extensive studies of fossils and was especially drawn to monstrous creatures.[38] Of the catalog of books attributed to him as author, few were published during his lifetime. They remained manuscripts passed around in copies for the edification of his friends and fellow naturalists.

Aldrovandi's monster epic is *Monstrorum Historia* (1642). Assembled by others from his manuscript materials and his collection of specially commissioned artwork, the book's arrangement follows that of a traditional cabinet of curiosities, starting with un-deformed man at the top. There follows a rogue's gallery of monstrous races: humans with huge ears that trail to the ground (*Homo fanesius auritus*), others with huge lips, or four eyes parade by as does the werewolf-like, dog-headed cynocephali. It then moves on to conventional monstrous births. Rendered in striking detail individuals with multiple arms and heads or no arms or no legs, and conjoined twins of several types appear. These include two separate examples of conjoined twins of a man and a wolf-man similar to the cynocephali. Wild men and women covered in hair or going all fours, which may have been an inspiration the later English poet and artist William Blake. Like so many other books and pamphlets on monsters the Ravenna creature and Monk Calf are in attendance.

Having covered man, the book goes on to include animals, birds, fish, and plants. Aldrovandi had little interest in living things that did not exhibit deformity. He preferred monsters of any kind over known living things, the few exceptions being deformed or abnormal individuals, such as dogs with missing front limbs. The book includes several birds like the bizarre and many plumed *Gallus Indicus*, and the even more outlandish rooster *Gallus Monstrosus* with its lion-like tail. In the section on marine animals there is a pair of amorous merpeople, the *Monstra Niliaca Parei*. Page after page of frightening creatures slouch, trundle and swim past in the extensively illustrated tome. Even the plants shown give off an otherworldly air. Included with the life drawings the text includes several drawings of skeletons of birds and a few mammals similar to the one Pierre Belon included in his work, though not arranged in a comparative fashion.

For naturalists like Gesner and Aldrovandi, the study of monsters, "wonders," and "novelties" did not undermine their worldviews but caused them to modify their thinking to account for such oddities in order to make their cosmologies whole and rational. If monsters were real, they had to be accounted for. Monsters also helped bridge the gap between the learned and the lay population. Intellectual distance could be surmounted because the public's appetite for monsters was as powerful as that of scholars. Monsters were a commonality that could work to the benefit of the elite virtuosi, particularly in acquiring examples. The scholars were mostly book and library-bound academics. Farmers, workers, and rural people lived in the field and were much more likely to encounter a monstrous creature than university professors and wealthy collectors. This worked to Aldrovandi's benefit, but it also left him open to skullduggery.[39]

Aldrovandi built up an impressive collection which became well known not only to other collectors, but to the public who were allowed to view it. He was one of the first monster collectors. His reputation as an erudite collector of choice objects and examples of strange beasts was such that when a dragon was discovered, it made its way to his hands. On May 13, 1572 a dead "dragon" came to light outside Bologna, and Aldrovandi soon acquired it. The creature caused a sensation amongst local people and word spread. Requests from other virtuosi to see the creature and to get copies of drawings of it flooded in.[40] Aldrovandi commissioned an artist to produce drawings of the dragon, and then copies to be sent out to other collectors.

Convinced the dragon to be genuine, and a portent of good luck for the recently elected Pope, Aldrovandi put together a manuscript on draconology which was circulated but not published until it was posthumously in 1640 as *Serpentum et Draconum*. The book is carefully arranged into two parts as per the title. Each creature or group of creatures has sections on its mythology, mysticism and allegory, and even numismatics and toxicity where applicable. The drawings, while artistically impressive show what can only be a sign the artists did not see the actual specimens. Snakes, like in Gesner, have bulging heads with distinctive horns. There is a series of drawings of real-life marine rays and skates altered to look more monstrous. Aldrovandi's dragons, at least the ones that appear in his book are copied from Gesner, and as stated previously, Gesner's go back to the later Middle Ages. The leathery wings, bipedal stance, scaly, somewhat tubby body (as if a snake had just swallowed a rabbit) dominated iconography. The epitome of this design, what Gesner called *De Serpentibus Alates* is Aldrovandi's *Draco Aetheiopicus*.[41] Whatever Aldrovandi's dragon may have actually been, it likely did not look like the illustration used in the book.

Tales of monsters and dragons like Aldrovandi's influenced literary authors as well as natural historians. By 1603 Shakespeare could make reference to a "monster in his thought, too hideous to be shown" in *Othello* without any need to explain. Giambattista Basile's *Lo Mercante* (1634) draws upon scientific monster texts, particularly writings about the "Venetian Hydra," in his story of dragons and dragon slayers. Conrad Gesner rejected the Venetian Hydra as a hoax in his *Historia Animalium* yet showed an illustration of it twice. He pointed out how profitable and widespread was the industry of manufacturing fake monsters in Venice. As Basile was going for a literary effect, whether the hydra was genuine was irrelevant, it suited his storytelling purpose.[42]

The Renaissance and early Modern naturalist collectors saw examples of the deformed as having as much or even more to say about living things than the finely made.[43] Known animals were easy to confirm and describe as actual examples could be found in Europe or brought back from far off lands to be placed in the menageries that began to appear across the continent. The aims of the scholar virtuosi collectors like Aldrovandi included clearing up questions of some of the odder and more fantastic legends of animals. The effort to determine the validity – what would be called debunking today – of mythical creatures took up a good bit of the effort. Most, if not all, the Renaissance virtuosi pursued both human and animal monsters partly to learn more of the nature of living things, but also out of a simple desire to collect and own the most extravagant creatures possible. A Wunderkammer might contain as many oddities as examples of traditional natural objects. This desire pushed along zoological understanding as much as anything else they did. It helped them move away from blind acceptance of authorities and toward a more modern understanding of the acquisition of zoological knowledge from direct observation.

Yet at the same time questions were being asked about the nature of monsters and whether they really were vehicles by which to study nature. In 1560, Florentine author Benedetto Varchi (1502–1565) argued that monsters were mistakes and errors of nature. Taking the opposite view of many of his contemporaries, these mistakes, he said, should be avoided by the naturalist in favor of the non-wondrous. Monsters, he believed, trampled on nature's rules and so should be avoided by those studying the natural world. "Monsters" of any kind repulsed Varchi whether in politics, the arts, or nature.[44] The bulk of naturalist authors, however, continued to pursue such creatures.

The New Virtuosi

In the Middle Ages, monstrosities were considered bad omens: things to cause concern and fear. Malformed animals and especially humans were the most prized by collectors. They upset what was considered the proper balance of existence. A two-headed baby, for example, was a sign of God's wrath, and a warning of perils to come. By the 16th century, while many of the poor and uneducated classes still shuddered with horror at the appearance of a malformed child, the new Virtuosi saw them as windows on the complex state of existence.

Collectors were not just collectors anymore; they had become curators. The collections were organized not just by type of object but arranged in such a way to tell a story. The cabinet of curiosity Virtuosos helped begin the Scientific Revolution. In the 19th century, a new cadre of virtuosi came along. Wealthy aristocrats and royalty still held their collections, but now medical doctors began creating teaching collections that contained monsters and that rivalled anything organized by the traditional virtuosi. One such medical collector was Rudolf Virchow (1821–1902). A complex character who held Liberal political views and participated in the German Revolution of 1848, Virchow was expelled from medical school. In 1849 he was hired by the University of Würzburg to teach pathology. He eventually found a seat in the Reichstag. He saw medicine as a form of politics. He produced an astonishing amount of work, so much so that some thought there were several people publishing under one name. Despite his liberal political position he held strong anti-Catholic views. He famously called Charles Darwin, whose work was just starting to be known, an ignoramus, and though he did not disagree with the idea of evolution he saw no evidence for it. He rejected the Aryan theory, argued the recently discovered Neanderthal was a deformed modern human rather than an ancient one, and invented the term *Kulture Kampf* (Culture War). Virchow's medical research led him to coin the term Leukemia, and he was a pioneer of public health.[45]

Virchow also understood the value and importance of teaching his students how to recognize disease and how to heal injuries. He realized the best way to do that was to show students examples of such things while still in school rather than wait for them to appear in their clinics. He amassed a large teaching collection which included monstrous births and many other types of medically related specimens. He, along with several others, collected medical specimens not for the shock value but as teaching aids to further the cause of medicine. He created a teaching collection in Berlin which would have made Rudolf II jealous and which outdid most of the traditional virtuosi.[46]

In America Thomas Dent Mütter (1811–1859) trained as a physician and surgeon at the University of Pennsylvania and then Paris. While overseas he came into contact with European medical cabinet collectors. As the Vroliks did in Amsterdam, and Virchow did in Germany, Mütter put together a private collection of anatomical specimens – especially unusual ones – and then donated them to the College of Physicians of Philadelphia in 1856. By the 1870s, the college, so convinced of the efficacy of teaching with such a collection, acquired the one belonging to Viennese anatomist Joseph Hyrtl and others. This became the Mütter Museum in Philadelphia. The collection focused, as many such teaching collections did, on anomalous human deformities and other examples of medical oddities and injuries. The compelling idea behind the medical teaching collections was that it was a useful way to show medical students what a condition looked like so that they were familiar with it. This allowed more effective diagnoses. Along with diseased anatomical parts, injuries were also collected. Snapped bones, infected eyes, tumors, and other such items became highly sought after.[47] Eventually, the Mutter Museum contained monstrous births, and in the great tradition of P. T. Barnum, they had a full body cast of Chang and Eng Bunker, the original Siamese twins.

End of an Era

The last of the great monster collectors in the pure virtuoso tradition of the early modern period was the Italian Fortunio Liceti (1577–1657). Holding a doctorate in philosophy and a medical degree, Liceti's *De monstrorum natura, caussis, et differentiis libri duo* (1616) dealt with biological malformation on the embryonic level.[48] Eschewing theological explanations he saw monstrous births of either animals or humans as natural. He tried to explain deformities through hereditary causes rather than divine or demonic ones. Liceti argued that while conjoined twins, for example, were rare but natural phenomena, human-hybrids were not. Deformities and monsters appeared, he argued, as a result of nature trying to work its way around some natural biological impediment. Liceti wondered about the nature of the word monster. Rather than *monstrum* meaning portent, he asked if it came from *monstrare* or "to show." He did not think deformities were curses from God or "errors" by nature, but nature's attempt to adapt and be better. A two-headed baby being the result of not enough material for two or too much for one.[49] Along with the text Liceti included a number of drawings of monstrous births. His depiction of a Cyclops baby (born with a single eye in its forehead) is thought to be the first ever published and based on an original drawing by none other than Leonardo da Vinci.[50] Liceti's leaning toward biology and away from theology presaged work by others.

With the end of the Tudor period and the advent of the Baroque in the early 1600s, the Roman author Pliny, while still a powerful force in natural history and monster studies, had begun to go out of favor with naturalists. The virtuosi Humanist tradition of relying on authorial texts was giving ground to the empiricist philosophy of direct observation and original sources. Greater knowledge of animals and people outside Europe helped bring on the slow steady fade of the monstrous races as well as some of the more outrageous creatures like dragons. There were, however, still beings lurking about which gave naturalist authors reason to pause thanks to the legendary baggage and veil of mystery which clung tenaciously to them. The literature on monsters produced during this period in books, articles, pamphlets, and broadsides is prodigious, so much so it is beyond the scope of this work to do a complete review of it here. What is important here is that monsters lurked about everywhere in the scholarly and popular imagination as well as on the printed page to the point of being overwhelming. They had become a legitimate focus of scientific study but still inspired theologians and apocalypticists. This vast body of work was indicative of the high level of interest in strange bodies, monstrous creatures and the myths associated with them. During this period, however, some grew tired of mythology altogether.

The 17th century saw Francis Bacon (1561–1626) articulate the modern scientific method (inspired again by Islamic alchemists) which rejected reliance upon the type of past authorities the collectors were enamored of. Interestingly, Bacon did not reject monster studies. He argued that scholars should engage with the types of monsters and biological wonders the collectors focused on. In the *New Organon* (1620) Bacon also wondered about speciation. While he thought one species transmuting into another "would be very difficult," he thought it would be much easier proposition to vary the characteristics within a type. Either way, to better understand the process by which transmutation in any form could take place, how "errors" of nature were produced and explained, it was important for naturalists to study monsters of all kinds. Toward this end Bacon said, "we have to make a collection or particular natural history of all prodigies and monstrous births of nature; of everything in short that is in nature new, rare, and unusual." It had to be done in a rigorous and scholarly manner to get the best results. Superstition should be avoided, as should "natural magic and alchemy."[51]

In 1662, one of the last major monster compendiums appeared. Written by German Jesuit Gaspar Schott (1608–1666) it repeated material already well known with little original thinking or work. The *Physica Curiosa Sive Mirabilia Naturae et Artis* included the *Gallus Monstrosus*, the bizarre rooster along with cut-away drawings of eggs with human heads with snake-like appendages inside. Schott also included drawings of primates: a "satyri" looking like Mr. Tumnus from Narnia, cynocephali, a human with an elephant's head, as well as various hairy men and women. In the monsterologist tradition, Schott took many of the illustrations he used from other authors including Edward Topsell. Primarily an engineer who wrote on hydraulics, he was mentored by Anastasias Kircher and kept up with the latest advances in the growing scientific revolution, corresponding with Robert Boyle.

The tradition of the collector naturalist combining the religious, mythical, and secular gave way with the work of the English naturalist John Ray (1627–1705) who stripped his scientific writing of any poetry or literary references. Primarily a botanist, Ray, collected specimens in the field so as to study plants directly rather than by reputation. In 1667 he was elected a fellow of the Royal Society. He traveled across England and Europe collecting plants which he used as the basis of his *Historia Plantarum* (1686). He tried to define the term species and was an early advocate of the idea that fossils represented the remains of once living organisms.[52] While not a monster enthusiast, Ray helped draw natural history studies further away from the medieval reliance upon past authorities and legendary descriptions and to a more modern engagement based upon direct observation and study of actual living things. He disagreed with many of his colleagues over the reality of spontaneous generation, arguing it was not occurring, "as hath been the constant opinion of naturalists heretofore."[53] He worked on experiments to show that insects formed around dead organic material for natural reason rather than spontaneous or anomalous reasons.

Despite their efforts to medicalize and naturalize monsters, the collectors and scientific virtuosi fought an uphill battle. As they published their increasingly scientific and scholarly work, authors like Gesner, Paré, Aldrovandi and others had to share space with popular broadsheets and pamphlets still pouring off presses in some profusion. The sensational interest in monsters continued as the naturalists looked more scholarly. Cheap pamphlets and broadsheets showed no sign of giving way. If anything, the production of the popular publications on monsters grew more vigorous. More of these tabloid style works saw print, purchase and perusal than the expensive lavishly illustrated science books discussed here.

A more practical approach to monsters came in the form of Giambattista dela Porta's *Magiae Naturales* (1558). An Italian scholar, Dela Porta (1535–1615), also known as John Baptista Porta, studied with tutors and on his own became a noted botanist and gifted cryptographer. He wrote on human anatomy and had assembled a cabinet of curiosities so impressive it inspired the prodigious author Athanasius Kircher (1602–1680) to create one in Rome.

His fascination with natural history led him to form a science club known as the Otiosi in order to investigate the secrets of nature. However, this led some – including the Inquisition – to suspect him of occult delvings. This, despite the fact that Dela Porta was a devout Catholic, and Jesuit lay brother.

Dela Porta's *Natural Magick* was on the surface, a household companion. It contained chapters on distilling wine, cooking, beauty tips for women, and some other interesting ideas. There are chapters on "Strange Glasses," "Strange Cures," and "Pneumatic Experiments." He also discussed the generation of plants and animals, including monsters. He had some ideas about the mutability of species. He thought almost any combination of creatures could produce "new kinds of living creatures," which "may be generated of diverse beasts by carnal copulation."[54] Insects out of mud, serpents out of a man's marrow, and scorpions out of basil were all possible, and all possible by average householders.

There was a deeper meaning to his text, however. Natural magic as an area of study strove to give naturalistic, non-metaphysical explanations to the occult forces of nature: forces most assumed were demonic. Techniques could be employed to tap into natural magic to manipulate nature for some required end and that doing so did not constitute dabbling in witchcraft or sorcery. When one engaged in natural magic one was simply using the forces of nature that God had supplied in abundance. Nature could produce odd blends of animals and humans thus monsters occurred naturally and not as the result of demonic influence.

The Church opposed the use or study of natural magic in part because it wanted to retain control over the supernatural. Church authorities wanted to have a monopoly on the metaphysical and natural magic suggested average people, if they knew the techniques, could access these forces on their own. Dela Porta strove to make natural magic an empirical science done in part through experimentation. As a result, Dela Porta found himself being investigated by the Inquisition and accusations of sorcery made against him.

Sorcery seemed to follow monsters. There had been a rivalry between the influential French author Jean Bodin (1529–1596) and the German physician and natural philosopher Johann Wier (1515–1588) over the reality of witchcraft. Wier argued that the witch trials then going on were wrong headed because those accused of witchcraft had not done anything truly demonic and had not practiced sorcery. They had, he said, only thought they had done such things through the use of various natural drugs and concoctions. Foolish women, he said, had not flown on broomsticks but had applied a special salve to their skin which made them think they had flown. They were not Satan worshipers, only sad, lonely, deluded individuals. Bodin staunchly believed, as many scholars did, that witchcraft existed, that witches did fly on brooms – and all the other things they were accused of – and had cavorted with demons, and that monstrous creatures were real.

To prove his point, Wier used Dela Porta's *Natural Magic* as a source book. Dela Porta explained how to make the salve that caused flying hallucinations. Bodin then accused Dela Porta of being a sorcerer for having spread this information. The Inquisition was already suspicious of Dela Porta for his involvement in the *Secreti* science society. Fortunately for Dela Porta, he was able to explain himself sufficiently to the authorities and he managed to escape punishment. He argued that showing how such salves were made helped eliminate superstition, not promote it. He also argued that monsters were natural as well and not to be feared as the products of the devil. He was as upset with the superstitions of the scholarly world as that of the popular. He also shot back that it was Bodin who was the heretic for promoting the idea that sorcery and witchcraft existed when it did not.[55]

Humans too could produce monsters. Dela Porta had a low opinion of his fellow men arguing that they would copulate with almost any other creature "for man is of all other creatures most lecherous." All this carnal activity which he says "I am much ashamed to speak of," leads to "producing monstrous and half-savage broods" of men. This included half-man, half-goats. These hybrids will "set upon virgins, and by force ravish them."[56]

The Cabinet Rearranged

The virtuosi collectors made progress in the organization and understanding of the living world. Their cabinets, however, for the most part entertained on a circumspect and elite level. By the early part of the 18th century, the concept of the private wunderkammer evolved yet again. Now they found employment for more public enlightenment as teaching collections in hospitals and universities. Medical schools across Europe and later North America assembled medical cabinets to train students. At Paris, London, Leiden, Amsterdam and Berlin forward looking professors of medicine and anatomy followed the earlier virtuosi in looking to deformities and monsters as vehicles to understand generation and heredity. Physicians such as Petrus Camper (1722–1789), Frederik Ruysch (1638–1731), and Gerard Vrolik (1775–1859) made Amsterdam a center of medical teaching cabinet use. These men utilized their collections of anatomical parts and monstrous births to train future doctors to recognize various medical conditions. Camper studied the relationship between humans and apes. Where earlier collectors had produced art to illustrate their work these researchers advanced ways of taking organic specimens and preserving them. Ruysch along with colleague Jan Swammerdam experimented with ways to inject tissue with wax to preserve them as anatomical models. Vrolik's son Willem Vrolick followed in his father's footsteps training as a naturalist and physician. Of his publications *Handbook of Pathological Anatomy* (1827) focused on congenital deformities like cyclopia and conjoined twins. He saw nature as a window upon the divine, but that nature had an innate internal life force pushing it along. Monstrous birth defects could show the life force running amok. Vrolik saw humans as part of nature not above it. The close seeming relationship between groups – particularly humans and primates – did not suggest transmutation to him however: he couldn't quite take that last step to transmutation.[57]

As the great age of virtuosi collectors evolved into modern museums, the center of monster collecting moved from the chateaus and castles to universities and museums. Where such collections were originally meant to impress the elite and their dilletante friends, monsters and deformity became teaching tools to benefit all. There was still the chance that a person could encounter an actual monster.

Notes

1 For the historiography of the Scientific Revolution see; Lorraine Daston, "The Nature of Nature in Early Modern Europe," *Configurations* 6, no. 2 (1998): 149–72, and Steven Shapin, *The Scientific Revolution* (University of Chicago Press, 1998).
2 E. W. Gudger, "The Five Great Naturalists of the Sixteenth Century: Belon, Rondelet, Salviani, Gesner and Aldrovandi: A Chapter in the History of Ichthyology," *Isis* 22, no. 1 (December 1934): 21–40.
3 Helen Woodruff, "The Physiologus of Bern: A Survival of Alexandrian Style in a Ninth Century Manuscript," *The Art Bulletin* 12, no. 3 (September 1930): 226–53. For the age of the *Physiologus* see, Alan Scott, "The Date of the Physiologus," *Vigiliae Christianae* 52, no. 4 (November 1998): 430–41.
4 Michael Curley, ed., *Physiologus* (University of Chicago Press, 2009).
5 Curley, *Physiologus*, 11–13.
6 Martin Luther and Phillip Melanchthon, *The Papal Ass of Rome and the Monk Calf of Freyberg* (London, 1579).
7 Luther and Melanchthon, *The Papal Ass of Rome and the Monk Calf of Freyberg*, 357.
8 Julie Crawford, *Marvelous Protestantism: Monstrous Births in Post-Reformation England* (Johns Hopkins University Press, 2005), 30–5.
9 F. D. Hoeniger and J. F. M. Hoeniger, *The Development of Natural History in Tudor England* (Folger Books, 1979).
10 See Liana Saif, *The Arabic Influences on Wearly Modern Occult Philosophy* (Palgrave Historical Studies in Withcraft and Magic, Palgrave Macmillan, 2015), and Mark Waddell, *Magic, Science and Religion in Early Modern Europe* (Cambridge: Cambridge University Press, 2021).
11 James W. P. Campbell, *The Library: A World History* (University of Chicago Press, 2013).
12 Patrick Mauriès, *Cabinets of Curiosities* (London: Thames & Hudson, 2002). Also see: Bruce Janacek, "A Virtuoso's History: Antiquarianism and the Transmission of Knowledge in the Alchemical Studies of Elias Ashmole," *Journal of the History of Ideas* 69, no. 3 (July 2008): 395–417.
13 Mauriès, *Cabinets of Curiosities*; Janacek, "A Virtuoso's History."
14 Paula Findlen, *Possessing Nature: Museums, Collecting, and Scientific Culture in Early Modern Italy* (University of California Press, 1996), 3.
15 Katherine Park and Lorraine Daston, *Wonders and the Order of Nature: 1150–1750* (Zone Books, 1998).
16 Jacob Rueff, *De Coneptu et Generatione Hominis* (Zurich, 1554).

17 A. W. Bates, *Emblematic Monsters: Unnatural Conceptions and Deformed Births in Early Modern Europe* (Rodopi: Amsterdam, 2005), 75.
18 Francis Dickie, "Ambroise Paré: A Barber's Apprentice Who Became the Greatest Surgeon in History," *American Journal of Nursing* 31, no. 10 (October 1931): 1143–6.
19 See Stephen Paget, *Ambroise Paré and His Times, 1510–1590* (G.P. Putnam's Sons, 1897).
20 Francis R. Packard, *The Life and Times of Ambroise Paré* (Paul B. Hoeber, 1921).
21 Bates, *Emblematic Monsters*, 75.
22 Bates, *Emblematic Monsters*, 75.
23 Ambroise Pare, *Des Monstres et Prodiges*, trans. Janis L. Pallister (University of Chicago Press, 1982), 3.
24 Katherine Park and Lorraine Daston, "Unnatural Conceptions: The Study of Monsters in Sixteenth and Seventeenth-Century France and England," *Past & Present* 92 (August 1981): 20–54.
25 Paré, *Des Monstres et Prodiges*, 12.
26 Monica Barsi, *L'énigme de la chronique de Pierre Belon* (Milano: Edizioni universitarie di lettere economia diritto, 2001), and Hoeniger and Hoeniger, *The Development of Natural History in Tudor England*, 49.
27 *Conrad Gesner: Physician, Scholar, Scientist, 1516–1565: A Quatercentenary Exhibit* (MD: National Library of Medicine, 1965), and Willy Ley, *Conrad Gesner: Leben und Werk* (Verlag der Münchner Drucke, 1929), also Hoeniger and Hoeniger, *The Development of Natural History in Tudor England*, 43.
28 Alfredo Serrai, *Conrad Gesner* (Rome: Bulzoni, 1991).
29 Bestiary, Therouanne? ca.1270, JP Getty Museum, Ms. Ludwig XV 3, fol. 89.
30 Stephan Batman, *The Doome Warning All Men to the Iudgemente Wherein Are Contayned for the Most Parte All the Straunge Prodigies Hapned in the Worlde, with Diuers . . . Conuersion Towardes God* (London, 1581).
31 Batman, *The Doome Warning All Men*, 93.
32 In 1658 both books were reissued together as *Historie of Foure-Footed Beastes and Serpents*.
33 Edward Topsell, *The Historie of Serpents* (London, 1608), 160.
34 Edward Topsell, *Historie of Foure-Footed Beastes* (London, 1607), 2–19.
35 Virgil B. Heltzel, "Some New Light on Edward Topsell," *Huntington Library Quarterly* 1, no. 2 (January 1938): 199–202.
36 Michael Bath, "The Sources of John Abbott's Pattern Book," *Architectural History* 41 (1998): 49–66.
37 Gian Battista Vai and William Cavazza, eds., *Four Centuries of the Word Geology: Ulisse Aldrovandi 1603 in Bologna* (Bologna: Minerva, 2003).
38 Andrea Baucon, "Ulisse Aldrovandi (1522–1605): The Study of Trace Fossils during the Renaissance," *Ichnos* 16, no. 4 (October 2009): 245–56.
39 Findlen, *Possessing Nature*.
40 The creature was a fake likely made from the remains of a sea going ray. Stanley Finger and Marco Piccolino, *The Shocking History of Electric Fishes: From Ancient Epochs to the Birth of Modern Neurophysiology* (Oxford: Oxford University Press, 2011), 108.
41 The images discussed here appear on pages 404–423 of *Serpentum et Draconum*. This figure from Aldrovandi also appears on a tapestry in the background of a scene on the fourth Harry Potter movie.
42 Suzanne Magnanini, "Foils and Fakes: The Hydra in Giambattista Basile's Dragon-Slayer Tale, 'Lo Mercante'," *Marvels & Tales: Journal of Fairy-Tale Studies* 19, no. 2 (2005): 167–96.
43 David Williams, *Deformed Discourse: The Function of the Monster in Mediaeval Thought and Literature* (McGill Queens University Press, 1999).

44 Park and Daston, *Wonders and the Order of Nature*, 201–2.
45 Thomas Glick, *A Comparative Reception of Darwinism* (Chicago: University of Chicago Press, 1988), 86–7, and L. J. Rather, "Rudolf Virchow: Standpoints in Scientific Medicine," *Bulletin of the History of Medicine* 30, no. 5 (September–October 1956): 436–49, J. P. Machenback, "Politics Is Nothing But Medicine at a Larger Scale: Reflections on Public Health's Biggest Idea," *Journal of Epidemiology and Community Health* 63, no. 3 (March 2009): 181–4.
46 Unfortunately, a large part of Virchow's collection was incinerated during the Battle of Berlin during World War II.
47 Gretchen Worden, *Mütter Museum of the College of Physicians of Philadelphia* (New York: Blast Books, 2002).
48 Fortunio Liceti, *De Monstrorum Natura Causis et Differentiis* (1616).
49 Elena Lazzarini, "Wonderful Creatures: Early Modern Perceptions of Deformed Bodies," *Oxford Art Journal* 34, no. 3 (2011): 415–31.
50 Armand Marie Leroi, *Mutants* (New York: Viking Press, 2003), 71.
51 Francis Bacon, *The New Organon: Or True Directions Concerning the Interpretation of Nature* (1620), chapter XXIX.
52 Charles Raven, *John Ray, Naturalist: His Life and Works* (Cambridge: Cambridge University Press, 1942).
53 John Ray, "The Extract of a Letter Written by Mr. John Ray to the Publisher from Midleton, July 3, 1671. Concerning Spontaneous Generation; As Also Some Insects Smelling of Musk," *Philosophical Transactions* 6 (1671): 2219–20.
54 John Baptista Porta, *Natural Magick* (London, 1658), 28.
55 William Eamon, *Science and the Secrets of Nature* (Princeton University Press, 1996).
56 Eamon, *Science and the Secrets of Nature*, 42–3, and *Natural Magic*, Chapter XII.
57 Laurens de Rooy, *Forces of Form: The Vrolik Museum* (Amsterdam: Vossiuspers Uva, 2009).

3

WORRISOME WOLVES AND BEASTLY MEN

Outside the small French village of Ubacs, the 14-year-old shepardess, Jeanne Boulet, was attacked by something. Blood and chunks of flesh were everywhere. It was June 30, 1764, and Jeanne Boulet was dead. What would come to be known as the Beast of Gévaudan had made its first recorded kill. Between 1764 and 1767, almost 200 people were attacked in the south-central French region of Gévaudan, with a hundred losing their lives. Terrified local people searched for the attacker, the authorities became involved with special hunters sent in, and even King Louis XV took notice. What was this creature or creatures that seemed to pass up animals to attack humans, preferring women and children? Was it a wolf or a wolf pack? Or was it, as some whispered darkly, a werewolf?[1]

For most of its history monster studies were largely an intellectual pursuit: the discourse on both human and animal monsters going on with a certain theoretical detachment. By the early modern period, however, that distance narrowed and monsters stepped up close and personal. They came to occupy the realm of the visceral and physical as well as the temporal. The search for werewolves and ape-men had immediacy to it. They were supposedly flesh and blood organisms that tore at the fabric of conventional wisdom as well as its flesh. They could be encountered not only at cultural boundaries, but in everyday situations, and they might just kill you.

The Strange Lives of Werewolves

Unlike any of the creatures discussed so far, the monster of the werewolf held special fascination and dread. Monstrous births, no matter how horrifying, were thought to pose a threat to the larger community, which would be the

DOI: 10.4324/9781003488767-4

recipient of generally bad times brought on by the appearance, for example, of a two headed baby. The werewolf did not bring on widespread calamities to society at large rather it posed a direct fatal calamity for individuals. For most of recorded history, but particularly by the early modern period, the half-man, half-wolf, lycanthrope reigned as a ferocious reality. It looked like a wolf and may have been part man.

Adding to the conundrum of werewolves, there was no physical evidence. Monstrous births could be examined and specimens acquired by naturalists and physicians. Werewolves, despite the widespread reports of their existence, left no physical traces. The most popular and oft-told stories of encounters with werewolves always resulted in the capture of fully human persons and never the recovery of a piece of a werewolf or a hybrid creature. As a result, investigators had nothing to study except reports. Even true believers could not come to any consensus as to what constituted a werewolf. Were they physical or metaphysical, figments of the imagination, creations of deranged minds, or flesh and blood organisms?

The werewolf is a creature that through various procedures can turn into or become a wolf. Sometimes the human becomes completely a wolf morphologically and sometimes a part human, part wolf monstrosity. There is little uniformity to the stories other than this.[2] The concept can trace its lineage back at least to 2000 BCE, with several variants appearing in Greek mythology. The poet Virgil tells of Moires who was a wolf-man and who learned magic from his wife the Goddess Moera. The most famous Greek werewolf story appears in the works of Ovid, Hesiod, and others, and concerns Lycaon. While several variants appear, the most popular story has Lycaon, the King of Arcadia preparing a meal of the flesh of a murdered child for Zeus in order to test his divinity. Outraged, Zeus turns Lycaon into a wolf – though often depicted as a wolf-man – for eternity. Pliny discussed how some believed werewolves to be genuine entities. He makes reference to men turned to wolves, relating a story in which a man swims across a body of water and is turned into a wolf for nine years. He sees these stories as false and adds with a bit of a jibe at his fellows how "It is astounding to what lengths Greek credulity will go; there is no lie so shameless as to lack a supporter."[3]

The first use of the term werewolf in the Medieval European context may have been in the *Ecclesiastic Ordinances* of the early British King Cnut (985–1035 CE), though this may have been a reference to the Biblical allegory of the sheep in wolf's clothing, or political imposter rather than a monstrous shape shifter.[4] The term Lycanthrope first appears in Reginald Scot's *Discourse on Witchcraft* (1584). He considered acting like a wolf a mental disorder using the older term *Lupina melancholia*. All along, scholars had wrestled with the idea of the nature of werewolfery: was it an actual transformation or only a psychological one? Did a scientific explanation for lycanthropy exist? How could it be explained rationally and naturalistically? While tracing its origins back to the classical world, and appearing in many different cultures, the heyday of the werewolf came in the seventeenth and eighteenth centuries in Europe. Wider Western views on werewolves came from a surprisingly small number of high-profile cases.

While the "werewolf" attacks in Gévaudan were eventually put down to attacks by hungry, conventional wolves, a case occurring a century and a half earlier did have a human protagonist. An adolescent 13–year-old named Jean Grenier was accused by witnesses of having engaged in unspeakable acts of murder, cannibalism, and werewolfery in and around the St. Sever district of Gascony. When interrogated, Jean claimed he had met an older man whom he called the "Lord of the Forest," who rubbed him with a special salve and wrapped him in a wolf hide which turned him into the animal. With this newfound power, Jean selected out children he found alone in the woods and attacked them. He would bite their throats suffocating them then drag their bodies to a secluded spot and devour them. He would never eat all the body leaving the leftovers for other wolves. The court believed he had murdered at least fourteen children in this fashion. Jean readily admitted all this insisting he was indeed a werewolf. Several witnesses, including a young girl confirmed most of his story. After considering the evidence the court decided that Jean was not in his right mind and must be suffering from some mental aberration. They did not believe he had actually transformed into a wolf-man – or at least in his case a wolf-boy. He seems to have been treated badly as a child by his family and employer and as such developed a warped view of life. Popular reaction to his crimes called for Jean's death along with his father Pierre. Because of the extenuating circumstances, rather than execute Jean he was confined in a monastery where he died seven years later.[5] There were several cases during the 16th and 17th centuries in France all of which have narratives similar to the Jean Grenier case, but his remained a popular one and became the most often referred to as an example.

The changing of a man into a wolf was not always seen as completely supernatural, some authors were almost nonchalant about the natural aspect of the transformation. In his *Natural Magic* (1558) Giambattista Della Porta explained, along with home remedies and cooking tips, how one could transform one species into another creating a new creature, at least a new monstrous creature. His work was popular, though briefly banned by the Inquisition. An English translation of 1658 said that to turn into a werewolf one need only apply a specially prepared salve. The book is full of his examples of how to create monsters from living things. Almost any combination could produce "new kinds of living creatures . . . generated of divers beasts, by carnal copulation."[6] He had a low opinion of his fellow men arguing that they would copulate with just about anything "for man is of all other creatures most lecherous."[7]

An unusual variant on the wolf-man model was the human-wolf monstrous birth. These were individuals where a wolf and a man were literally joined. A number of illustrations of such combinations appear in texts from the early modern period. In *De Monstris* (1665) Fortunio Liceti was so taken with the idea of conjoined twins of a man and a wolf he included two separate illustrations of them. The first and more crudely drawn is on the frontispiece of the book while the second more carefully rendered version appears later in the text.[8] These were not occult creations rather they were naturally appearing. "The werewolves are certayne [sic] sorcerers, not only unto the view of others they seem as wolves, but to their own thinking they have both the shape and nature of wolves."[9]

Modern Western scholarship as well as folklore sees the werewolf as a horrific menace associated with murder, witchcraft, and Satanism. This view came largely from French sources. Works such as Claude Prieur's *Dialogue de la lycanthropie* (1596), Jean Bodin's *De la lycanthropie* (1615), and Georg Newenfeldt's *Disquisitio philosophica* (1650) became leading source texts.[10] More recent scholarship has focused on works from Germany, Italy, Eastern Europe and the Baltic region. One of the historiographic issues to be confronted in the study of werewolves is the wide range of inconsistent lore that surrounds them. No werewolf equivalent to the witch's *Malleus Maleficarum* existed as a guide. Stories and legends of werewolves and werewolf activity go from standard depictions of them as merciless child killers to ones in which the werewolf is a friendly, even heroic figure recovering stolen goods and fighting against the forces of evil. Some stories tell of humans transmuting completely into wolves while others only wear wolf skins. Some involve the use of hallucinogenic potions while others do not. Rumanian stories for example, say that one can only turn into a werewolf at certain times of the year and that one can be born a werewolf if their parents engage in certain inappropriate behavior around Christian holidays.[11] Regions such as Hungary had no indigenous werewolf legends, those that appeared there were imported from Germany.[12] The ethnic and political landscape from which these stories emerge are complex labyrinths that make standardizing them or forming stereotypical models difficult. (Ironically, what might be called the modern view of what werewolves are comes not from these local legends but from 20th-century Anglo-American cinema and literature).

An example of this confused history is the odd case of Thies Kaltenbrun. Often referred to as the Livonian Werewolf, Thies was a local 80-year-old character, living outside the city of Riga in what is now modern Estonia and Latvia. He had a reputation as a curmudgeon and eccentric. In 1691, while giving testimony in a robbery case, he related (completely unsolicited) to the court that he was a werewolf and had traveled to Hell – in the form of a wolf – with some friends in order to retrieve grain and other foodstuffs hidden there. They transformed, he said, by donning wolf pelts. Thies, however, denied being a Satanist. In fact, while he readily admitted to being a werewolf he insisted he and his compatriots fought against the forces of evil and were the "hounds of God." Rather than punish Thies for werewolfery, the judges ordered him flogged for recruiting others to turn away from Christianity. That this was the crime he was punished for rather than werewolfery shows reluctance on the part of the judges to support belief in genuine lycanthropy. Some later scholars have argued for werewolves being shamans rather than shape shifters. Others have argued that there is no evidence that werewolves had any connection to a shamanistic tradition. The story of Thies is also problematic from a shamanistic point of view in that there is no evidence he performed that role. Historian Willem De Blécourt warns that in studying werewolf lore, "the danger of over interpretation lurks everywhere."[13] Stories like that of Thies Kaltenbrun call into question any normative legend of werewolves.

As the Enlightenment and Scientific Revolution dawned, a debate ensued over whether demons could transmute a human into a werewolf. Philosophers and theologians wondered whether the human soul was capable of becoming bestial or whether it was some type of mental affliction. It was during this period that psychological explanations for lycanthropy gained currency. These theological reservations posed the same problems for werewolfery that evolution did two and a half centuries later: was the condition a natural occurrence or a divine one.[14] The werewolf legend slowly began to dissipate as did belief in an actual creature. This decline increased due to competition from another man-like creature of folklore.

The Ape

The late 15th through mid-18th centuries saw an increase in world trade and extensive exploration. Advances in sailing technology as well as the opening of land routes such as the Silk Road brought increased knowledge of the outside world that was empirical rather than mythical. The pursuit of capitalism brought growth and change to economies. These events also had a direct impact on monster studies. Old myths of Blemyeas, sciopods, and werewolves went into decline, but other monsters were confirmed. Ships that traveled the seven seas brought not just tales of strange foreign animals, but the creatures themselves.

Monsters – whether animal or human – were the ultimate acquisition for any collector with a cabinet of curiosities. This stemmed largely from the perceived uniqueness monsters possessed. This uniqueness made them sought after subjects for study as well as for collectability. It also made them exceedingly difficult to acquire. Actual specimens – aside from Aldrovandi's dragon for example – created special difficulties for understanding the nature of monsters. Monstrous births of humans and animals, while rare, could be acquired for study. Other legendary beasts proved harder to come by. The acquisition of apes led the way in understanding mythical and legendary creatures and proved a bit more accessible than dragons.

Some of the smaller primates had made it into Western menageries. A painting of Henry VIII and his family done in 1545 and hanging at Hampton Court, for example, shows the young Will Somers with a small tailed monkey playfully on his shoulder. Baboons, the Cynocephali, were known – though not on an anatomical level – as well. The larger anthropoid apes, however, remained clouded in legend. It was unclear if the stories told about "apes" were of truly unique monsters, some lower form of human, or just myths. While Greek scholars had some firsthand experience with primates, by the Middle Ages, European scholars had little chance of ever seeing one at home or abroad – except for the few examples of small-tailed monkeys that had been acquired by a few in the nobility. As such, naturalists did not focus much attention on them. There seemed to be so little interest in primates that Marco Polo's description of ape-like creatures in the land of Comari (likely India and Japan) generated barely any notice.

African people were long since acquainted with primates as were the people of Asia. Some early societies saw primates as humans who had failed some Divine test and thus transformed into beasts by the Gods or had been denied full humanness. The Cerops of Greek myth were men turned to apes by Zeus for tricking Hercules. The image of the ape or monkey as trickster lasted well into the modern era. The ape took on the role of a cowardly con artist. Christian theologians equated apes with dark skinned brutish enemies of Jesus, thus anyone not of the correct religious persuasion – heretics, Jews, Moslems, and others – were depicted as conniving monkeys in Christian tradition. In the *Physiologus* monkeys were said to represent "the very person of the Devil."[15] Aside from these broad racial stereotypes, little was known of the biology of primates by European scholars (little of an anatomical nature of primates was known even in cultures where they were common). Legends attributed to apes included them attacking men, and especially women, and generally brutish man-like behavior. Increasingly, however, naturalists wanted facts not fairy tales. A new era in zoological writing was about to dawn, one based upon empirical knowledge of animal anatomy. One of the earliest depictions of a man-like (anthropoid) ape came in the herbal *Hortus Sanitatus* (1491). Simply drawn, a cynocephalus, werewolf-like "Cephos" stands bipedaly and carrying a stick. Sometimes attributed to Johann von Cube this may be the first example of the "monkey on a stick" artistic motif that would come to dominate depictions of primates well into the 19th century.

In 1667, the German Jesuit intellectual and author Athanasius Kircher (1601–1680) released his *China Illustrata* and discussed primates. Kircher's interests ranged wide and included having one of the more compelling cabinets of curiosity of his day. A polymath who studied and wrote on theology, engineering, zoology, linguistics and geography, Kircher wanted to travel to Asia as a missionary. Instead, he wrote his book based upon the reminiscences and experiences of Jesuits who had traveled there.[16] He relates how in China can be found "a hairy animal very similar to a human." It has a voice like a man, runs very fast and "when it meets a man on the road it smiles like a human then attacks him."[17] Kircher assumes these are apes rather than man-like monsters of myth. He makes the distinction between these creatures and wild men (feral people raised by animals) who are genuine humans. In his *Arca Noe* (1675), he suggested that animals could adapt to a new environment and change. He used the example of deer released from Noah's ark that encountered cold climates and changed into reindeer. He also said that some species were hybrid crossings of others. Kircher's traveler's observations on these aggressive primates were brief. He had no direct experience with them: few Europeans did and the ones that did had little anatomical knowledge of them. Not until a ship from Africa docked in England in 1697 did Western naturalists get their first look at the man-like monsters which haunted so many travelers' tales.[18]

Primates had become disconcerting to some, but in morphological rather than theological terms. There simply was very little actual scientific knowledge of primate anatomy. Several species still found themselves referred to by the same names used by Aristotle and Pliny: "cynocephalus" for the Baboon and "satyr" for the Orang-Outang. Europeans tended to lump primate groups together. The term Orang-Outang covered any primate found in the tropical zones of the world. This was a variation on the Malay term for "man of the woods," the local name for the creature. The Latin term *Homo sylvestris* reflected the notion, common to both South East Asian people as well as Europeans, that the Orang-Outang was a human, albeit a dark-skinned, degenerate form. The Dutch anatomist Nicholaeus Tulp (1593–1674) first used the term in print for a chapter titled "Homo Sylvestris, or Man of the Woods" in his larger *Observationes Medicae* (1641).[19]

Nicolaeus Tulp – who changed his name from Claes Pieterszoon because he thought his patients who could not read would be better able to remember his name and the giant tulip sign hanging outside his office – studied medicine at Leiden and began a long career as a successful physician, professor of anatomy, and local politician in Amsterdam. His work as an anatomical demonstrator inspired Rembrandt to make him the center of the famous painting *The Anatomy Lesson of Dr. Nicolaeus Tulp* (1632).[20] In 1641 he published his *Observationes Medicae* ostensibly as a guide to proper medical technique, based upon his own experiences. He included a number of discussions of the unusual animal life being brought to Holland by ships of the Dutch East India Company.[21] One of these creatures was a humanoid ape. Possibly the first such creature ever brought to Europe Tulp also called it a "satyrus indicus."[22] Despite this name, the animal was from Angola, Africa rather than Southeast

Asia. He considered the creature so important he used an illustration of the animal as the book's frontispiece. This drawing also found its way into the 1665 edition of Liceti's *De monstrorum natura* and turned up reprinted in a number of later texts. Tulp saw the creature at a menagerie, but did not get a chance to examine it in close detail or dissect it. As a result, his text is by necessity limited. With these shortcomings, his text stands as the first tentative modern attempt to clean up some of the mythology that had grown around the primates, and bring them into the fledgling biological sciences as reality rather than legend. While Tulp wrestled with the work of past authorities, he could not bring himself to ignore them completely. "Never can we retreat from the opinions which have acquired age," he said, "or from fables invented by poets."[23] After Tulp the major leap forward in what would become primatology appeared.

In 1678, the Polish-Scottish naturalist Joannes Jonstonus (1603–1675) published *A Description of the Nature of Four-Footed Beasts*. Not to be confused with Edward Topsell's book of the same name, Jonstonus – also known as John Jonston – was a physician and naturalist whose father had immigrated to Poland from Scotland and had connections to local aristocrats. He studied at several universities in Europe as well as Cambridge. Unlike many of the zoology authors of this period, Jonstonus was not particularly interested in monsters. He did, however, discuss primates. Rather than the religious parables or fables of other authors he relates a story of how "a noble man in England kept [a primate] that kept his rabbits from weasels."[24] Jonston paid special attention to the ape's anatomy, including discussing the structure of their hearts. He never sees the apes as monsters or as a deformed version of man. He clearly articulates that the cynocephali are Baboons not men. He had never dissected or closely examined an ape himself, but he mentions how the Roman physician Galen had and makes reference to the work of Tulp and how he referred to Orangutans as wild men. Unlike Jonston, many 17th century naturalists mixed notions of theology and science in their examinations of monsters. They may have embraced the new science, but they did it with their arms wrapped in earlier garments. Something like a clean break, or at least an acknowledgment of making a break with the old mythologies, had to be made. A detailed anatomical study of primates would soon appear and help put to rest rumors about them and about monsters in general that had festered for centuries.

Edward Tyson

The serious study of primates, and indeed ape-men, begins with the work of Edward Tyson. A physician rather than a naturalist, Edward Tyson (1650–1708) was born in Bristol, England. His father was a mayor and an ancestor had traveled to the West Indies. He attended the University of Oxford, entering Magdalen Hall (today Hertford College) in 1667.[25] At Oxford, Tyson studied with early archaeologist Robert Plot (1640–1696) who was fascinated by natural curiosities. In his *Natural History of Oxford-Shire* (1677) Plot discussed some of the first fossils found in England. In *Natural History of Stratford-Shire* (1686), he argued that archaic stone tools found there had been produced by humans of the early modern period rather than the ancient world. Plot thought a large fossil femur he found outside Oxford must have come from a giant monstrous creature of some kind. He originally thought it came from an elephant, but he couldn't fathom how elephants might have made it into the British Isles. Along with studying with Plot, Tyson's student days were filled with dissections of humans and animals. After graduation, he moved to London to open a medical practice.[26]

In 1677 Tyson came into contact with the Royal Society and Robert Hooke, the polymath, microscope builder, architect and the first person regarded as a professional scientist.[27] The Royal Society's publication, the *Philosophical Transactions*, meant to be the vehicle to convey the important scientific work done by society members and others to the outside world, already had a history of printing articles and letters about monstrous births and other biological anomalies. The chief experimenter of the Royal Society, Robert Hooke (1635–1703) befriended Tyson despite the difference in their ages. The earliest recording of them meeting is from Hooke's diary for May 29, 1677.[28] Hooke was 42, Tyson 27, and their friendship lasted until the former's death. Hooke published his groundbreaking book on microscopy, *Micrographia*, in 1665 and was working on the rebuilding of London after the Great Fire. Hooke too helped make slightly known creatures more so. In his way, Hooke showed monsters to be real. The illustration of a flea that he included in the *Micrographia* showed an alien, armored creature for all the world a monster. People knew of fleas, but their anatomy and structure remained almost mythical. Tyson did something similar as had his friend and mentor Hooke.

In London Tyson continued his dissections by examining a recently deceased clergyman, patient of his. This allowed him to accurately describe the action of gall stones in the human body for the first time. Tyson also took a position of curator of experiments for the Royal Society in 1682 along with fellow physician Frederick Slare (1648–1727).[29] To help Hooke, who was falling behind in his curatorial duties, Tyson conducted and organized demonstrations for a brief period.[30]

Tyson felt strongly that anatomical studies were crucial to successful medical practice. How could, he argued, a physician or naturalist say they knew the bodies of men or animals if they had never studied the inner workings of living things? To this end he helped pioneer the modern biological descriptive monograph, a publication dedicated to the investigation of a single specimen. His first major work of this type, *Phocaena, or the Anatomy of the Porpess dissected at Gresham College*, came in 1680. Having acquired a freshly deceased porpoise, Tyson dissected it in minute detail then described it in an equally detailed text with accompanying technical illustrations done by his "most ingenious friend Mr. Hook[e]."[31] Hooke was a skilled dissector and draftsman making drawings of specimens for the porpoise book (or "porpess" as Tyson called it).[32] Using his meticulous dissections and comparisons, Tyson showed the morphology of the porpoise was more like that of a land quadruped mammal than a fish as was commonly held. Just because it looked like a fish, he argued, did not mean it was a fish. He was helping to define the concept of homology, where similar structures in different species had different functions.[33]

Rather than simply describe the specimen, Tyson placed it in context as well. He consciously saw his age as one in that took knowledge of the universe far beyond what had been previously known. "New tracts, new lands, new seas," he said, "are daily found out, and fresh descriptions of unknown countreys [sic]" are brought in, "so that we are forced to alter our maps and make anew the geography of both again." Old superstitions, legends, and twice-told tales about strange animals and stranger humans were giving way to empirical study and science. This allowed myth to be replaced with fact. He and others like him no longer took "the fabulous records of the ancients" at face value as past naturalists had.[34]

Along with the careful descriptions of anatomy, Tyson speculated on the relational aspects of the porpoise as well. There was, Tyson argued, a string of connections between species. Not going so far as to postulate an evolutionary paradigm, it seemed logical that instead of abrupt differences between species there was a subtle gradation. There is a scale, he said, connecting worms and men, a link between them. The porpoise stood as "a signal example of the same [link] between land-quadrupeds and fishes."[35] The format of Tyson's *Phocaena* became the model he would use in subsequent publications, including the *Anatomy of an Opossum* (1698) as well as his most famous work.

Anatomy of a Pygmie

Always on the lookout for new creatures to study, in 1698 Tyson acquired the body of a "pygmie." In actuality, a chimpanzee, the creature had been brought by sailors to Bristol from Angola as a pet. It became ill on the trip, and so by the time it arrived in England it was close to death. Tyson saw the creature alive, but sickly, at least once. After its death, he acquired it and along with his friend, the anatomist William Cowper (1666–1709), dissected it. Cowper was a specialist in musculature and so brought that expertise to the examination of the body. He was also a skilled artist. It is Cowper who did the drawings of the pygmie that appeared in the book that was the result of this work (The Flemish artist Michael van der Gucht (1660–1725) created the engravings for the published version). Cowper also mounted the skeleton of the diminutive creature that still stands in the Natural History Museum in London.[36]

While doing the dissection, Tyson noted that his subject was a "remarkable creature, which not only in the outward shape of the body, but likewise in the structure of many of the inward parts, so nearly resembles man . . ."[37] As he worked the similarities to human anatomy struck him. He began to think that the legends of pygmies had originated with descriptions of primates. As a result, his book meant not just to show the anatomy of this particular specimen, but to show that pygmies and humans were not the same, pygmies were not human-like monsters, but a species all their own. He wanted to show the mythology to be wrong. "If I can do this," he says, "I shall think my time not wholly lost, nor the trouble altogether useless that I have had in this enquiry."[38] He then proceeds to lay out what ancient sources say about theses monsters and argues a direct examination of the actual anatomy shows these tales to be incorrect.

When the dissection was done, Tyson presented his findings to a meeting of the Royal Society on June 1, 1698. He brought along the skin of the animal, and may have had Cowper's drawings with him. The Royal Society audience was quite taken with the presentation and saw the resemblance to humans the pygmie exhibited. Tyson emphasized this resemblance. He also seems convinced that though he never saw the creature walk in such a fashion, it was bipedal. The morphology of the creature seemed clear to him: it walked upright. It was only on all fours when he saw it because, he said, it was so ill. In his interpretation of the Pygmie, Tyson came perilously close to articulating an evolutionary relationship between primates and humans, but not quite. He did see living things – even very superficially different things – as biologically related and saw the Pygmie as an intermediary between primates and humans. Later naturalists including Charles Darwin and Richard Owen acknowledged this. Tyson's biographer, Ashley Montagu, felt he had made a case for the doctor from Bristol as a pioneer of evolution as well as primatology. Montagu argues that "nothing could be clearer, no more definite."[39] On June 7 or 8, 1699 Tyson's work was released as *Anatomy of a Pygmie* and monster studies would never be the same.

Tyson chose the word Pygmie for a reason. His intent, as he explained in the introduction, was twofold. He wanted to show as explicitly as possible the anatomy of this creature and something else. Amongst naturalists and writers outside of Africa, knowledge of primates consisted of myth and legend supported by few empirical facts. Ideas about them ranged from being misshapen monsters, freaks, to some sort of degenerate human stock. Tyson wanted to end the speculation once and for all. Along with the carefully detailed anatomical information, Tyson's work sought to take a fictive creature and make it real. He wanted to show that "the Pygmies of the ancients were a sort of ape, and not of the human race," something he would "endeavor to prove in the following essay."[40]

Standing next to such creatures as Cynocephali, Blemmyea, and other monstrous races, the legend of the pygmie goes back to the *pygmaioi* of Homer.

These were diminutive creatures from Africa who often fought with cranes over food. For Tyson, the problem with descriptions of pygmies suffered because "the first account we have of them, was from a poet, so that they were only a creature of the brain, produced by a warm and wanton imagination, and that they never had any existence or habitation elsewhere."[41] Tyson's careful dissection of the actual animals might prove useful in determining where these creatures fit into the overall scheme of living things. "We may observe," he wrote, "nature's gradations in the formation of animal bodies and the transition made from one another." While scholars have argued over whether Tyson was thinking about transmutation, he clearly thought that the examination of monsters would contribute to such questions.[42]

In his philological essay, Tyson approached his study of the pygmie as an antiquarian/historian rather than as an anatomist/biologist. Modern historical scholarship began with the antiquarians of the 17th and 18th centuries. They sought to find verification of old legends, the origins of cities, cultures, artifacts, and kings.[43] Tyson approached the history of the pygmie legend in the same way. Rather than pushing away leaves and vines to read an ancient inscription directly as antiquarians would do to clear up questions about past civilizations, he pushed aside the superficial skin and morphology of the pygmie to read firsthand the animal's anatomy to better understand its place not only in biology but in history as well. Like antiquarians who wrote chorographies of specific places or cities, Tyson wrote a chorography of an individual animal: something he had already done with the porpoise. He addressed animal arcana as a primary source on past lives.[44] With *Anatomy of a Pygmie* Tyson introduced a new type of writing to the biological sciences: works which focused upon the anatomy of a single species but done by comparing them to related species. The concept of comparative anatomy he helped devise became a powerful analytic tool in understanding the structure of living things and their relationships to other living things.

Beginning his text, Tyson makes it clear that his reason for producing this work on the Pygmie concerns his desire to dispel various mythical beliefs of creatures he felt were misidentifications of existing, non-monstrous, species. In this way, Tulp (who harbored similar intent) and Tyson's work can be seen as an early cryptozoological enterprise: they strove to replace mythical creatures with genuine ones, to take the mythical and make it rational and scientific. Paradoxically, however, Tyson undermined his own effort to separate the monkeys with the life rendering he included of the creature. He depicted it in a bipedal stance supporting itself on what is clearly a man's walking stick. As a result, the image strikes the viewer as a hairy, beast-like human.

Tyson was sure the Pygmie (the term chimpanzee did not come into use until the later part of the 1700s) was a bipedal primate rather than a human who could not speak. Tyson was conflicted about the relationship between primates and humans. The morphological structure of apes was so similar to humans – something his work had helped establish – that he needed to find some way to separate them. He set upon language: humans had it while apes and pygmies did not. The difference was accounted for by Divine intervention.[45]

Modern historical research finds its origins in the antiquarians of the 18th century. Tyson's project of sorting out real from mythical creatures was as much antiquarian and historical as it was anatomical and biological.[46] Antiquarians in general but those interested in monsters like Conrad Gesner and Ulisse Aldrovandi united philosophical research (library and book based) with archaeological research (embracing physical artifacts). While not an antiquarian as such, Tyson came out of that tradition and was adding biological investigation in the form of anatomical dissections to the traditional antiquarian search for knowledge about the universe.[47] He took obscure occult-like anatomical lore and turned it into fact-based knowledge. His work in this area extended to a number of different anomalies.

Tyson was fascinated by unusual and malformed people and animals. He had drawings made of a woman with a huge goiter, as well as one of a cat and a human male both of whom had parasitic twins.[48] Like Darwin did a century and a half later, Tyson saw monstrous births as vehicles to examine human generation. To this end he personally went to observe a number of them. In 1688, a surgeon named Layfield called Tyson to view a child "very monstrous and surprising." The child had been born without a brain, its skull being deeply depressed down to the brow ridge. Though now deceased, the child was said by the midwife to have been born alive. Tyson examined the body then interviewed the mother. She told him that she had received a "considerable bruise in her belly" while pregnant. Tyson wondered if injuries to the mother could account for at least some irregular births. If the mother sustained an injury to her mid-section while carrying the child could the injury pass to the fetus?[49] Sensationalist pamphlets on monstrous births still found print and still played the divine retribution angle. In 1645, in England alone one was published every other month. Tyson discounted any notion of supernatural causes for birth defects, an idea not widely held even within the medical establishment of the day. He focused upon establishing prosaic, biological and medical explanations for abnormal creatures whether human or animal.

In May of 1699 – as *Anatomy of a Pygmie* was going to press – Tyson examined a monstrous birth of a pig with a human face. Those who gazed in wonder upon the creature all assumed it had been "produced by the copulation of two species." It was not uncommon to accuse someone of bestial behavior when confronted by such a deformed animal. After close scrutiny of the body, Tyson argued that some pressure upon the bones of the hapless pig's face had pushed it into a simulacrum of a human face. "I was convinced," he said, "that this monster was not from a conjunction of both kinds, but only occasioned by the pressure of the compression of the womb." Though he tried, he could not find out if the mother pig had been injured prior to the pig's birth.[50] Animals like this helped convince Tyson that species did not have the capability of cross breeding, certainly not with humans. It can be argued that the fixation on monsters by early modern scholars stemmed in part from the desire to challenge natural kinds, to undermine or support the idea of the fixity of species. An important natural history interest concerned whether species could cross hybridize and thus then could species really be fixed or could they change somehow over time? The eternal oneness of living things could be breeched, and not just breeched, but torn asunder. This would call into question many metaphysical and accepted truths about creation. Such an idea was ready to be accepted by some while others were vehemently opposed to it.

Monkey on a Stick

As the debate over the extent of the relationship between humans and apes continued "apes" in a generic form took on a sinister appearance as the dark and brutal other. Europeans and Americans equated them with people of color rather than themselves. Monsters – regardless of the culture that believes in them – take on analogous roles, usually depicting something unsavory. In Euro-American culture, the ape shouldered fears of race, gender, empire and all the malignant things lurking in the human psyche. Tyson's use of the name Pygmie and his discussion of mythical animals made a temporal connection between the Satyrs of classical myth, the Asian Orang-Outang, and the European Wild Man, as accusations of assaulting humans, especially women, hung around all their necks as they did for werewolves.

Tyson also contributed to the idea of the missing link by arguing that his Pygmie fell somewhere between apes and humans on the Great Chain of Being. He concluded that while humans and apes had superficial similarities, they ultimately fell into different groups anatomically and metaphysically. He did not want to upset Christian sensibilities and so strove to separate the two groups. Later contemporaries like Swedish systematist Carolus Linnaeus tried to push the two back together again with varying levels of success. Naturalists of the 18th century, in their efforts to work out the relationship between apes and humans, still acted within a Christian world view and not an evolutionary one. They tended to believe in the fixity of species. This meant that not only the non-human primates, but people of color occupied lower rungs on the ladder leading to God with Europeans at the top closest to the deity. Evolution assaulted this notion by making all men equal regardless of superficial differences and making them relatives, if not descendants of the apes.

The image Tyson used of the Pygmie, standing up like a biped balancing itself on what is clearly a man-made walking stick, had a long pedigree. First appearing as a crude drawing in Bernard von Breydenbach's *Peregrinatio in Terram Sanctam* (1486) where it was claimed the drawing was done from life. A version then appeared in the *Hortus Sanitatus* (1491) the monkey-on-a-stick was picked up by many other naturalists in their books. It appears in the *Viaggo de Venutia al Sancto Sepulchro* (1500), then in Conrad Gesner's *Historia Animalum* (1551) where it influenced Ullisse Aldrovandi in his *De Quadrupedibus* (1637). Jacob Brontius used a version in *Historia Naturalis* (1658) before Tyson used it in 1699. Tyson's sophisticated version reinvigorated it, where it was then used by Count de Buffon in his *Histoire Naturelle* (1749–1788) and a number of others. At about the same time, under the supervision of Linnaeus, the Russian naturalist Christian Emanuel Hoppius used it for *Anthropomorpha* (1760), and even an American version appeared in *Bickerstaff's Boston Almanac* (1785) which acknowledged Tyson as the source of its own crude version.[51]

The similarities between primates and humans grew difficult to ignore and helped diminish the werewolf as a man-like reality. They helped bring the two groups closer philosophically as well as anatomically. In his *Discourse on the Origin and Foundations of Inequality Among Men* (1755) Jean-Jacque Rousseau (1712–1778) famously discussed the Orang-Outang as a variation of "natural man." He suggested they were primitive humans who had yet to acquire the various technologies, behaviors, and vices of non-primitive Europeans. He drew on contemporary scientific sources such as Antoine-Francois l'abbé Prevost's *Histoire Générale des Voyages* (1746) who had built upon the work of Nicholaeus Tulp, George Louis le clerc de Buffon's *Histoire Naturelle* (1749), and others. Rousseau was not doing this in an attempt to support the concept of human evolution: his use of the Orang-Outang was more political than scientific. By placing the Orang within the human family, he could critique European social order and the divine right of kings.[52]

Conclusion

The growing knowledge of primate anatomy made mythical creatures like the werewolf increasingly problematic as an actual living organism. It also helped lead some to engage with ideas about transmutation, which questioned the fixity of species. If werewolves were real and they did transform themselves – though by now few learned persons believed the idea – was this a metaphysical effect or a biological one? Could permanent shape shifting occur, where one species transformed into another? While intellectually dismissing werewolves philosophers could not do that so easily with the apes. Some naturalists such as the Frenchman Jean-Baptiste de Lamarck, and English authors such as Robert Chambers, T.H. Huxley, Charles Darwin and others saw primate anatomy as support for a general theory of evolution, but more specifically of human evolution as well. Lycanthropy had been relegated to the realm of mental illness, and the actual creation of werewolves had been brushed aside as mythology, pushed aside in part by the ape. The departure of the werewolf from the realm of the real allowed an open space which would eventually be filled by large primates such as the Yeti, and Sasquatch, and Bigfoot. Science helped banish one monster while simultaneously allowing for the rise of another. As real creatures, however, the werewolves would always have their defenders. Werewolves and other beast men only existed as mythology. There were monsters a bit more existential. While still elusive, sea monsters were easier to see.

Notes

1 The story of the Beast of Gévaudan has been analyzed from the time of the attacks themselves. An excellent recent foray into the field is Jay M. Smith, *Monsters of the Gévaudan: The Making of a Beast* (Harvard University Press, 2011).
2 Most of what modern people think of as werewolf lore or 'facts' comes not from history, but nineteenth and twentieth century fiction and cinema.
3 Pliny, *Natural History*, vol. 3 (Loeb Library, 1940), 59. Also see, Richard Henry Crum, "The Werewolf Again," *Classical Weekly* xxvi, no. 13 (January 23, 1933).
4 Charlotte Otten, *A Lycanthropy Reader* (Syracuse University Press, 1986), 5–8. Dorothy Whitelock, ed., *English Historical Documents: 500–1042* (Routledge, 1979), 462.
5 Robert Eisler, *Man Into Wolf – An Anthropological Interpretation of Sadism, Masochism, and Lycanthropy* (Routledge and Paul, 1948) and Pierre De L'Ancre, *Tableu de l'Incostance de Mauvais Anges et Demons* (Paris, 1612).
6 John Baptista Porta (Giambattista della Porta), *Natural Magic* (London, 1658), 28.
7 Della Porta, *Natural Magic*, 43.
8 Fortunio Liceti, *De Monstris* (Amsterdam, 1665), 182.
9 Richard Verstegen, *A Restitution of Decayed Intelligence* (London: John Norton, 1634).
10 Claude Prieur, *Dialogue de la lycanthropie, ou, Transformation d'hommes en loups* (Louuain, 1596); Jean Bodin, *De la lycanthropie, transformation, et extase des sociers, auec la refutation des argumens que Bodin allegue au 6. chap du second livre de sa Demonomanie, pour soustenir la realité de ceste pretenduë transformation d'hommes en bestes* (Par, 1615), and Georg Newenfeldt, *Disquisitio Philosophica, Vol I De Oraculis Gentilium, Vol II De Lycanthropia* (Jenae, 1650).
11 Harry Senn, "Romanian Werewolves: Seasons, Ritual, Cycles," *Folklore* 93, no. 2 (1983): 206–15.
12 Gábor Klaniczay, trans., Susan Singerman, ed., and Karen Margolis, *Uses of Supernatural Power: The Transformation of Popular Religion in Medieval and Early Modern Europe* (Cambridge Polity, 1990), 160–1.
13 Willem De Blécourt, "A Journey to Hell: Reconsidering the Livonian 'Werewolf'," *Magic, Ritual, and Witchcraft* 2, no. 1 (Summer 2007): 49–67. Quote from page 66.
14 Nicole Jacques-Lefèvre, "Such an Impure, Cruel and Savage Beast," in *Werewolves, Witches and Wandering Spirits: Traditional Belief and Folklore in Early Modern Europe*, ed. Katheryn A. Edwards (Kirksville: Truman State University Press, 2002) and Brian J. Frost, *The Essential Guide to Werewolf Literature* (Madison: University of Wisconsin Press, 2003).

15 Curley, *Physiologus*, 39.
16 For Kircher's life see; John Glassie, *A Man of Misconceptions: The Life of an Eccentric in an Age of Change* (Riverhead Hardcover, 2012) and Paula Findlan, *Athanasius Kircher: The Last Man Who Knew Everything* (New York: Routledge, 2004).
17 Athanasius Kircher, *China Illustrata with Sacred and Secular Monuments, Various Spectacles of Nature and Art and Other Memorabilia* (Amsterdam, 1667) trans. Charles D. Van Tuyl (Indiana University Research, 1987), quotes from page 186.
18 The classics of ape history are H. W. Janson, *Ape and Ape Lore in the Middle Ages and Renaissance* (London: Warburg Institute, 1952) and William McDermott, *The Ape in Antiquity* (Johns Hopkins University Studies in Archaeology 27, 1938).
19 Nicholaas Tulp, *Observationes Medicae* (1641).
20 Robert M. Goldwyn, "Nicolaas Tulp (1593–1674)," *Medical History* 5, no. 3 (1961): 270–6, and S. A. Mellick, "Dr Nicolaes Tulp of Amsterdam, 1593–1674: Anatomist and Doctor of Medicine," *Australian and New Zealand Journal of Surgery* 77, no. 12 (December 2007): 1102–9.
21 Charles G. Gross, *A Hole in the Head: More Tales in the History of Neuroscience* (MIT Press, 2009), 172, and S. A. C. Dudock, *Nicolaeus Tulp: The Life and Work of an Amsterdam Physician and Magistrate in the 17th Century* (Six Art Promotion Bv., 1998).
22 Tulp, *Observationes Medicae*, 271.
23 Tulp, *Observationes Medicae*, 273.
24 English translation of John Jonston, *A Description of the Nature of Four-Footed Beasts* (London, 1678), 75. Translation by 'JP.'
25 William Tyson Papers, Jeffries Collection, Bristol Public Library, UK, SR42, volume 1–4, also see *Munk's Roll of the College of Physicians* (1861): volume 1:399–401.
26 The major biography of Tyson is Ashley Montagu, *Edward Tyson, M.D., F.R.S., 1650–1708* (Philadelphia: American Philosophical Society, 1943). Also see, Malcolm H. V. Cameron, "Seventeenth Century Science with Particular Reference to Comparative Anatomy," *Canadian Journal of Comparative Medicine* 11, no. 5 (May 1947): 132–7.
27 For Hooke's life see Lisa Jardine, *The Curious Life of Robert Hooke* (Harper Perennial, 2005).
28 Harry W. Robinson and Walter Adams, eds., *The Diary of Robert Hooke, 1627–1680* (London, 1935), 293.
29 Felicity Henderson, "Unpublished Material from the Memorandum Book of Robert Hooke, Guildhall Library MS1758," *Notes and Records of the Royal Society* 61 (2007): 129–75.
30 Stephen Pumfrey, "Who Did the Work? Experimental Philosophers and Public Demonstrators in Augustan England," *British Journal for the History of Science* 28 (1995): 131–56, 141.
31 Edward Tyson, *Phocaena, or the Anatomy of the Porpess Dissected at Gresham College* (London, 1680), 10.
32 Margaret Espinasse, *Robert Hooke* (London: William Heinemann, 1956), 138.
33 Lawrence Kruger, "Edward Tyson's 1860 Account of the 'Porpess' Brain and Its Place in the History of Comparative Neurology," *Journal of the History of Neurosciences* 12, no. 4 (2003): 339–49. Charles Newman, "Edward Tyson," *British Medical Journal* (October 11, 1975): 96–7.
34 Kruger, "Edward Tyson's 1860 Account of the 'Porpess' Brain"; Newman, "Edward Tyson," 1.
35 Kruger, "Edward Tyson's 1860 Account of the 'Porpess' Brain"; Newman, "Edward Tyson," 11.

36 Tyson kept the Pygmie skeleton in his possession and handed it down to his granddaughter who put it in the Cheltenham Museum until the early 1890s when it was moved to London. Bertram Windle introduction to Tyson's *Pygmies of the Ancients* (London: David Natt, 1894), xii.
37 Tyson, *Anatomy of the Pygmie*, 3.
38 Tyson, *Anatomy of the Pygmie*, 6.
39 Montagu, *Edward Tyson*, 244.
40 Tyson, introduction to *Anatomy of a Pygmie*.
41 Tyson, introduction to *Anatomy of a Pygmie*.
42 Jonathan Marks, "Contemporary Bio-Anthropology: Where the Trailing Edge of Anthropology Meets the Leading Edge of Bioethics," *Anthropology Today* 18, no. 4 (August 2002): 3–7. Also see, J. E. H. Smith, "Language, Bipedalism, and the Mind-Body Problem in Edward Tyson's Orang-Outang (1699)," *Intellectual History Review* 17, no. 3 (2007): 291–304.
43 Arnaldo Mamigliano, "Ancient History and the Antiquarian," *Journal of the Warburg and Courtauld Institute* 13 (1950): 286.
44 Bruce Janacek, "A Virtuoso's History: Antiquarianism and the Transmission of Knowledge in the Alchemical Studies of Elias Ashmole," *Journal of the History of Ideas* 69, no. 3 (July 2008): 400.
45 Smith, "Language, Bipedalism, and the Mind-Body Problem."
46 Mamigliano, "Ancient History and the Antiquarian," 285–315.
47 Bruce Janacek, "A Virtuoso's History: Antiquarianism and the Transmission of Knowledge in the Alchemical Studies of Elias Ashmole," *Journal of the History of Ideas* 69, no. 3 (July 2008): 395–417.
48 *Volume of Sketches by Dr. Edward Tyson*, Royal College of Physicians, MS618, page 10, 39, and 79 respectively.
49 Edward Tyson, "An Observation of an Infant Where the Brain Was Depressed into the Hollow of the Vertebrae of the Neck," *Philosophical Transactions of the Royal Society* 19 (1695–1697): 533–5.
50 Edward Tyson, "A Relation of Two Monstrous Pigs with the Resemblance of Human Faces," *Philosophical Transactions of the Royal Society* 21 (1699): 431–5.
51 See, Giacomo Giacobini y Riccarda Giraudi, "Y el hombre encontró al simio," *Elementos* 68, no. 14 (Octubre–Noviembre 2007), Página 3. In his *History of Anthropology, Volume 1*, (Routledge, 1996), 94 Frank Spencer argues that Hoppius only translated Linnaeus' manuscript into Latin for publication, and did not actually write it himself. *Bickerstaff's Boston Almanac* (Boston, 1785).
52 Francis Moran III, "Of Pongos and Men: 'Orang-Outang' in Rousseau's 'Discourse on Inequality'," *Review of Politics* 57, no. 4 (Autumn 1995): 641–64.

4

THE FEARLESS SEA-SERPENT KILLERS

The learned engagement with the monstrous found a new focus in the 18th and 19th centuries with things in the sea. It would also find its demise there. After millennia of being considered a natural part of the study of life on Earth, monsters of all types found themselves increasingly seen as an intellectual dead-end, as something unscientific, fit only for folklorists and historians to study as a way of gauging popular credulity in any given age. Monsters found themselves in an unusual position. As with any chart the peak of scientific interest in monsters, particularly sea monsters, in the Victorian Age, heralded its rapid and steep drop-off as well.[1]

Since humans first gazed upon large bodies of water, whether big rivers, lakes or oceans, they worried what might lurk beneath the surface. That there were genuinely odd, large, and dangerous creatures living in even shallow watercourses gave impetus to the creation of legends and myths about horrifying creatures that swam. For scholars, sea monsters were not just bookish and philosophical; they were genuine creatures made more real by reports of encounters with them from responsible and reputable sources. This gave justification to some naturalists to take them seriously, rather than dismiss them out of hand, stories of such unusual creatures.[2]

Upon closer examination most sea monster stories turned out to be misidentifications of known animals such as fishes and whales, while some defied explanation. Lending verisimilitude to the reality of sea monsters, the fossil record burst forth with a host of strange creatures that once swam the world's oceans. These remains proved that monstrous creatures did exist, at least in the past. In an odd twist it is with the discovery of the fossil marine reptiles that the lexicon of monsters became embedded in science and natural history studies at the same time that their bodies were being rejected. Complicating

DOI: 10.4324/9781003488767-5

the situation, scholarly research showed that along with fossils, at least one mythical monster tradition did have its basis in reality.

The Kraken

References in Western culture to water born monsters appear as early as the Old Testament. In the book of Job and others, mention is made of the leviathan, of dragons, and of crooked serpents of the sea.[3] They appear in Babylonian, Egyptian, Indian, and Chinese mythology. As already seen, Homer, Ovid, Aristotle, Hesiod, Herodotus, Pliny and other Greco-Roman authors discuss them.[4] Some early accounts of sea monsters are now thought to have been of genuine organisms such as whales, dolphins, and others not well known at the time. Legends of the most famous historical sea monster, the Kraken, fall into this category. Tales of the Kraken have been told in Northern European folklore for centuries. While there is some contention about authorship, a manuscript from around 1000 AD, said to be by King Sverre Sigurdsson of Norway (1151–1202?), is thought to be the first known textural appearance of the word Kraken.[5] Described in some myths, such as the 13th century *Övar-Odds Saga*, as being many armed and deadly. Due to its frightening tentacles the Kraken made a popular subject for illustrations and other visual representations, usually in the act of attacking a ship or hapless sailor.[6]

The Danish author and historian Olaus Magnus (1490–1557) wrote descriptions of the Kraken and transcribed several older stories associated with it in his *Historia de Gentibus Septentrionalibus* (1555). Erik Pontoppidan (1698–1764), bishop of Bergen, described it in his *Natural History of Norway* (1752) where the story reached a wide audience and became the source for many later works. There is also Hans Egede's (1686–1758) 1741 account in *A Description of Greenland*. The theologian and minister to Copenhagen wrote of a creature forty feet tall with flippers like wings and which spouted water like a whale. Egede did not see the creature himself. He was retelling the tale of a sighting made by his son Poul Egede (1708–1789), in 1734, and who published his own account around the same time as his father.[7]

The identity of the Kraken was known enough that the Swedish taxonomist Carl von Linné (1707–1778), known as Linnaeus, included it in the first edition of his groundbreaking *Systema Naturae* (1735) as a cephalopod calling it *Microcosmus*. In his attempt to account for all types of living things on earth, Linnaeus felt it acceptable to place mythical creatures into his system of biological organization. Along with the Kraken he included a collection of monstrous and anomalous creatures he categorized as the *Paradoxa*. These included such reported creatures as the Hydra, the phoenix, and the draco (dragon). In later editions, Linnaeus changed his mind and dropped the Paradoxa, but kept the Kraken.[8]

French naturalist Pierre Denys de Montfort (1766–1820) added to the scientific knowledge, but also to the fabulous legend of the Kraken in his *Historie Naturalle Générale et Particulière des Mollusques* (1802). Acknowledging it to be a giant squid or *poulpe,* he also claimed he knew a story of a ship off Angola that had been attacked by one. With axes and sticks the crew called on St. Thomas to save them and managed to run the creature off. In honor of their survival, they had a votive picture created and installed in the church at St. Malo France. Author Richard Ellis suggests it was from this story that Jules Verne gained the inspiration for his novel *20,000 Leagues Under the Sea* though that inspiration may have come from another source discussed a little later.[9] De Montfort's imagination could get the better of him. He claimed that a squadron of British warships that disappeared in 1782 had been devoured by giant squids.[10] De Montfort's fantastical ideas aside, by the mid-19th century large squid and octopi were known to naturalists. Dead specimens, or at least parts of them, had been washing up on shore, collected and preserved. While few details of their lives were known, their morphology was being pieced together and described. Their ferocious beak and sucker claws survived after death and were studied closely.[11]

Zoologist and botanist Johannes Japetus Smith Steenstrup (1813–1897), building upon the work of Olaus Magnus and a description in the Danish *Annalar Björns á Skardsa* (1639), too suspected the Kraken was a giant squid. In 1857, he gave the creature its scientific name, *Architeuthis*. Steenstrup was an acknowledged expert on mollusks, cephalopods, and barnacles. As director of the Zoological Museum of the University of Copenhagen, he did early work on cuttlefish, helping make them scientifically known. He also corresponded with Charles Darwin on Cirripedia (barnacles) and contributed fossils specimens to the British naturalist's collection.

As a result of the growing knowledge of the anatomy of squid and octopuses, reports and legends of multi-armed monsters were put down as sightings of these creatures. As Kraken legends were likely based on encounters with genuine creatures, naturalists could make a proper description of it, bringing it out of the realm of fantasy and into the world of biology. A myth had become real. That, however, left the snake-like sea monsters unaccounted for creating something of an intellectual conundrum of the kind that had plagued monster researchers since the classical age: if a sea-serpent was not a squid or octopus they must be misidentifications, or they must be real, but scientifically unknown: they had to be something.

The American Sea Monster

Swimming monsters were not confined to Scandinavia. Sightings of sea-serpents and lake monsters appear in Native American lore as well as stories told by European explorers and settlers. The coastal Native people had a host of sea and lake monster stories. In 1899, anthropologist Albert S. Gatschet of the American Bureau of Ethnology, used the term "miraculous hydrozoology" to describe these stories, making him a proto-cryptozoologist of a sort.[12] A number of naturalists looked at North American monster sightings, trying to explain them. One of them was the French expatriate Constantine Samuel Rafinesque (1783-1840). One of the supremely unique individuals in the annals of natural history, Rafinesque drove himself relentlessly. He seemed to have wanted to classify every living thing on earth and expended considerable energy toward doing so. Not a professional in the modern sense of the word, he was a gifted amateur who feverishly worked for social respect and position by collecting and publishing (with only a brief stint with academic affiliation) but received little. This is not unusual. With a few exceptions, the majority of naturalists, and historians as well as archaeologists, in the early 19th century, were amateurs with little formal training in their field. What they lacked in academic credentials they made up for with a passionate zeal for studying the natural world (the word "scientist" would not be coined until the 1830s after which it still took some time before it was widely used).[13] Working largely by himself, he was a "lone voice in the wilderness."[14] Rather than any monetary concerns, workers such as Rafinesque were driven by a deep inner interest in figuring out how the universe worked.

Rafinesque was born in Constantinople but raised in France and Italy. A voracious appetite for knowledge steered him to become a prodigious autodidact of natural history. In 1815, he came to the United States traveling extensively, botanizing, and naturalizing amongst the vast array of living things of the New World. By 1819, he found a position as professor at Transylvania University at Lexington, Kentucky. He was critical of Linnaeus and strove to create and promote his own rules of classification. He published descriptions of living things at a frantic pace in a wide range of publications. He also had a tendency to openly disparage colleagues he did not care for. In retaliation, a damning eulogy by the American botanist Asa Gray (1810–1888) along with years of bridge burning, left Rafinesque's reputation as a naturalist in tatters. He was widely considered a crank well into the 20th century.[15] Despite his quirks and idiosyncrasies, however, Rafinesque did useful work on sea monsters as well as on evolution.

While Charles Darwin was just returning from the *Beagle* voyage, Rafinesque already argued for biological transmutation of species, including using the terms "mutation" and "spontaneous evolution." In 1836, both his *New Flora and Botany of North America* and the *Flora Telluriana Pars Prima* by Rafinesque made open arguments for evolution when few others dared. Based upon his study of the flora and fauna of North America, including sea monsters, he put forward the notion that species were new breeds of existing varieties and that in the past there had been fewer species than in the present, if not only one or two groups which became all the subsequent species.

Unfortunately, Rafinesque put forward no explanatory mechanism for how these evolutionary changes came about so his ideas were not picked up on. In later editions of *On the Origin of Species* Darwin made reference to a number of naturalists who had come close to explaining evolution including Rafinesque. In his "Historical Sketch," Darwin made direct reference to, and used quotes from, the *New Flora*. Other than that, Rafinesque received little acknowledgement for his work on biological change.[16]

Ever on the lookout for new and unusual organisms, Rafinesque turned his attention to something strange in New England. Sightings of mysterious sea creature, off the coast of Massachusetts went back to the 1600s. During the summer of 1817 a series of reports came in from eyewitnesses who claimed to have seen a serpent-like monster prowling the waters off Gloucester and Cape Ann, Massachusetts. Local fisherman and beach combers said they saw the creature while others claimed to have seen baby monsters in a nest on the beach. These sightings were so widely known and popular the artist Elihu Vedder produced the painting *Lair of the Sea-Serpent* based upon the Cape Ann sightings in 1864.[17] The media from Boston to New York and beyond contained eyewitness reports of what this creature might be. Various descriptions had it as snake-like, 40 to 50 feet in length, and sometimes said to resemble a string of floating barrels.[18]

Amidst the excitement, the Boston based, Linnaean Society of New England thought the stories presented an opportunity for American science. The society had begun as an elite gentleman's club catering to well to do men with natural history interests. However, they had a bit of an identity problem, not really knowing whether they wanted to stay private and restricted or establish a more public profile. Correspondents sent in reports and eyewitness testimony, which the society collated for publication. They engaged a local justice of the peace, former Massachusetts state senator, Lonson Nash, to collect depositions from witnesses. The product of all this activity was the *Report of a Committee of the Linnaean Society of New England Relative to a Large Marine Animal Supposed to be a Serpent* (1817). It was mostly made up of eyewitness stories rather than any scientific analysis of evidence. A farmer had killed a snake near the beach around the time of the sightings and it was thought this might be a baby monster. Society representatives examined the dead creature's remains and noticed some unusual aspects to its vertebra. Assuming the dead animal a juvenile form of the sea going beast, the society named the Gloucester monster *Scoliophis atlanticus*.

In addition to the society's pamphlet, Professor William Dandridge Peck of Harvard wrote of the creature in the *Memoirs of the Academy of Arts & Sciences* while his colleague Jacob Bigelow wrote an item for the *American Journal of Science*, edited by the pioneering naturalist Benjamin Silliman.[19] Peck (1763–1822) was a Boston medical doctor who became the first professor of natural history at Harvard while Bigelow (1787–1879) was another

Harvard doctor with an interest in botany. With the help of influential naturalist Samuel Mitchell (1841–1894), Peck debunked the offspring monster as a common local variety of snake. After a time, sightings dropped off, but the monster reappeared in 1818 and 1819. As late as 1947, it was supposedly sighted again off Gloucester.[20]

Few if any of the learned authors who wrote on the Cape Ann Sea monster went to look for themselves. One who did go, though he never actually saw the creature, was Revolutionary War hero, politician, naturalist, and fellow of the Royal Society of London, David Humphreys. He sent a series of letters on his adventure to Royal Society President Joseph Banks, for comment. Humphreys, "after several days fruitless search for the anomalous animal in question" analyzed what he had at hand.[21] To Humphreys, most of the eyewitnesses seemed the salt of the earth and so these sightings had to be taken seriously. He finished his letters to Banks by stating that the local revenue cutter was taking on extra guns in case they should encounter the thing they could blast it out of the water. His letters – quickly compiled into a book – while ostensibly a scientific treatise on a newly discovered creature are written in a literary way rather than a technical one which also included a plot line relating to American national pride. He saw the creature in terms of a monster from classical literature. He reified the Gloucester sea-serpent in order to learn moral lessons rather than biological ones, offering up great glory to whoever might capture or kill the beast. "Who will dare to put," he asks with a flourish, "a hook in his nostrils, transfix his sides with a spear, or engage in single combat with Vulcanean arms this monarch of the deep?"[22] He also thought that when captured the creature would show the grand design of God. The great serpent should remind people of the serpent in the Garden of Eden. Poetics and nationalist and religious ardor aside, Humphreys admitted the evidence was still a bit thin on the ground to prove conclusively, especially since the creature lives "in the dark and unfathomable abyss of ocean, among the unsearchable secrets of the GREAT DEEP [his capitols]."[23]

In his article, Jacob Bigelow rolled out a string of original letters and newspaper items to support the veracity of the accounts. These included a story from 1804 about an unnamed American warship that went looking for the monster. The captain tried to get his marines to fire on the creature, but before they could get ready, it had ducked under the waves never to be seen again. Bigelow also argued how could so many people at different times and at different places have described the same creature? He called sea monsters "the most interesting problem in the science of natural history."[24] Like Humphreys, Bigelow contacted Joseph Banks. While Banks did not reply to Humphreys' letters or book, Bigelow claimed that Banks had replied to him. By 1820, Banks was at the end of a long and distinguished career as a naturalist and was indeed infirm, and not overly eager to reply to American claims of monsters without attribution. However, Bigelow claimed Banks "expressed his full faith in the existence of our [American] Serpent of the Sea."[25]

At the same time, Constantine Rafinesque, hearing of such a strange creature, compiled information on his own then quickly published his analysis of the incident (he also did not personally go to New England to investigate). Ignoring how many American naturalists came out in favor of the creature Rafinesque biographer Leonard Warren says, somewhat incorrectly, that "American naturalists, ever commonsensical, dismissed these tales of the [Cape Ann, Gloucester Sea-Serpent] for the nonsense they were," yet Rafinesque weighed in.[26] His *Dissertation on Water Snakes, Sea Snakes and Sea-Serpents* sums up what was known about genuine aquatic snakes and compares that knowledge to what had been written about the Gloucester creature. Rafinesque begins the *Dissertation* on a skeptical note about sea monsters by stating that "a crowd of superficial writers hasten to offer us, instead of facts, their own conjectures on the subject," which he says are, "often wild, incorrect, or ridiculous." He explained away most sea monster sightings because "sea-serpents are merely sea snakes of a very large size." Just as the reader thinks he will be dismissive of the Gloucester sightings, he says that it is likely a giant unknown version of the known sea snake "belonging probably to the genus *Pelamus*." He then suggested it be given the scientific name of either *Pelamus megophias* or *Megophias monstrosus*.[27]

Rafinesque worried what others would think of the embrace of monsters by American naturalists. News of the creature soon reached Europe and the United Kingdom, and Rafinesque's worries were fulfilled. European and British naturalists sniffed at the Americans, particularly their descriptions of the creature and its habits. They took to task those who claimed the creature had come on land to lay its eggs. French naturalist Henri de Blainville thought the Linnaean Society of New England had erred considerably when they equated the dead snake on shore with the unproven monster of the waves.[28] Others saw implications for the understanding of biology in these sea creature stories. Reports of the Gloucester monster suggested the creature came on shore to breed, making it an amphibian rather than a reptile. From a biological point of view, this did not make sense and so the Americans must be wrong. An anonymous article in *Blackwood's Edinburgh Magazine* titled *On the History of the Great Sea-serpent* wondered about the descriptions of the creature's behavior. If such an amphibious serpent did exist it would have had to go through extensive biological change for "such changes never take place without corresponding alterations in the most important organs." If they did, they would undermine current knowledge of living things because "alterations of that nature have never been observed to occur in any of the snake tribe."[29]

Eventually the Linnaean Society of New England had to admit that the baby sea monster discovered on shore was a common snake. Worse yet, in 1818, a local ship Captain named Rich, fixated on discovering the source of the sightings once and for all, found the creature. To him the sightings had a familiar ring to them. He and his crew went out on calm days and discovered that the source of the sightings was a large, but common mackerel. During calm periods, these large fish swam close to the surface. This produced exactly the surface effects reported by eyewitnesses.[30] In 1861, sea monster author Philip Henry Gosse opined that he did not believe the Gloucester monster was real because Americans often make up tall tales and are unreliable when it comes to science.[31] Rather than bolster their reputation, the Linnaean Society's engagement with sea monsters undermined their credibility.

Despite the condescending attitude of some Europeans, a number of American naturalists thought the creature did exist. One was Benjamin Silliman (1799–1864). Yale educated, Silliman studied chemistry and natural history, became a professor at Yale, and rose to the ranks of one of the leading men of science in early republic America. He did work on the nature of meteorites, was an early member of the fledgling American Association for the Advancement of Science and edited the highly regarded *American Journal of Science*. American naturalists like Silliman felt like poor relations to their European counterparts. European savants generally regarded Americans as naive, impulsive, not as well educated as themselves, and generally producing inferior work. In addition, the Americas received mixed reactions from their own people. American science in the early decades of the 19th century had little respect or social status at home or abroad.[32] They wanted to change that and thought the Gloucester sea-serpent might be able to help.

Silliman and other American naturalists eager to prove their mettle, that they could do work on a par with Europe, felt proving the existence of a new and wholly unknown species would give them and their country greater scientific respectability. Silliman, however, based his conclusions not on any physical evidence of the monster, of which there was none, but on the veracity of the eyewitnesses. They were left with only ontological discussions of the nature of sea-serpents. None of the American naturalists who had come forward with opinions on the Cape Ann beast had ever seen it with their own eyes. Proving the veracity and reliability of the eyewitness reports was all these men had. This helps account for the rush to get a local magistrate to the scene in order to take official depositions from witnesses. They strove to build their case as if in a courtroom rather than a laboratory. This approach had been already used overseas as in the case of the Beast of Gévaudan. If eyewitnesses are beyond reproach, have no reason to lie, have nothing to gain from reporting seeing a monster, they must be telling the truth. It also can be used to mask the absence of physical facts. In 1879, a similar creature was reported off the coast of New Jersey near Sandy Hook. In 1894, something was reported prowling the waters of Lake Hopatcong. It was duly christened, Hoppie. Neither of these had any more evidence, other than eyewitness accounts, than the Gloucester monster.

In early republic-age America, this approach bolstered, and was drawn from, the nationalistic belief that good solid Americans were not liars or obfuscators, so if they said they saw a monster you'd better believe them. Sailors especially were sturdy folk, honest to a fault, but it was of no use. Europeans saw even learned American naturalists as bumpkins reporting monsters at every ripple in the waves. In England, sightings of sea monsters there were also supported by depositions rather than physical evidence, so several naturalists there came out in opposition to such techniques. It was reliance upon legal documents rather than physical evidence which helped to undermine belief in sea monsters by the growing mainstream scientific consensus which wanted the authority of university and museum-based empiricism to win out over hearsay.

While these sightings were taking place off the East Coast of America, real monsters were being found in Old England. One of the factors contributing to heightened interest in sea-serpents in the 19th century among naturalists, and what put off the final detachment of monsters from scientists if briefly, was that while no living examples had been brought in, strange and wondrous creatures began to be found in the fossil record. Quaint legends and descriptions from ancient texts, medieval manuscripts, and sailor's lore were one thing, but here were actual fossil remains of the most astounding creatures ever to come to light. These creatures had swum the oceans, as majestic and terrifying as anything from Greek mythology.

Mary Anning and the Plesiosaurs

Born in the Southwest English town of Lyme Regis into the family of a poor carpenter, Mary Anning (1799–1847) was taught to read and write at a local dissenter's Congregationalist Church. The pastor, Reverend James Weaton, promoted the idea to his parishioners that though God had created the world in a literal six days, it was the responsibility of good Christians to study the latest teachings from the new field of geology as well. Understanding the growing knowledge of the complex intricacies of the earth's structure would bring believers to a greater understanding of God's plan. Scientific discovery, therefore, was a boon to the righteous.

Rocks, minerals, and fossils could also be a source of income. Mary's father collected shells to sell to tourists at Lyme Regis. Taking her father's lead, she and her brother began exploring the rocky coast with picks and hammers. They collected their first fossil sea beast when she was only ten or twelve years old. She sold it to a local collector for the tidy sum of £23. Her mother joined them and a family business was born. As time went by, they found more examples, which excited a growing cadre of earth science enthusiasts. Along with geology, paleontology was in its infancy in England and the strange creatures Mary Anning and her family unearthed had never been seen before. She did not discover the first known examples of these fossils; however, several fragmentary specimens had been found previously by others but were thought to be the remains of crocodiles. The fossils Mary found were beautiful specimens often all but complete.[33]

Found in marine deposits and clearly swimmers of some kind, the fossils from Lyme Regis combined the characteristics of crocodiles, fish, and reptiles. Their flippers, long necks and tails put them more in tune with the monstrous denizens of mythology than any living creatures. The naturalist Everard Home commented on Mary's discoveries saying the creatures were "one of the connecting links in creation, formed for the purpose of preventing any void in the chain of imperceptible gradations, from one extreme of animated beings to another."[34] They showed that God made these creatures so that the great chain of being of all life on earth would be a smooth one, and not full of heretical question producing gaps. Home suggested it be called *Proteosaurus*.[35] Seeing they were not crocodiles, in 1818, the short-necked fossil monsters Mary Anning unearthed were dubbed *Ichthyosaurus* by naturalist Charles König. In 1821, the long-necked versions were labeled *Plesiosaurus* by Henry de la Beche and William Conybeare.[36] Practitioners of the new sciences were uncovering exciting new facts about the earth and its life making it a heady and romantic time for science and monsters.

Naturalist William Buckland, professor of geology at the University of Oxford, at first thought the plesiosaur might be a fake it was such an odd conglomeration of parts. Finally coming around to accepting the creatures as real he said of one particular specimen that it was "altogether the most monstrous," given its "neck of enormous length resembling the body of a serpent." Buckland, as did many of his colleagues, subscribed to the notion of Natural Theology, that the vast complexity of the natural world showed not some mechanical, natural process as was being talked about in some radical circles, but the designing hand of God.

Buckland (1784–1856) studied at Oxford and was especially attracted to the burgeoning field of geology. After he graduated, he spent time traveling across England studying first-hand the many unusual geological formations and fossil sights to be found there. In 1813, he was appointed Reader in Mineralogy at Oxford, and in 1819 the school created a new position – Reader in Geology – just for him. By 1824, he was president of the Geological Society of London and over the years held a number of positions within the Church of England. His position and many publications made him a major player in early 19th century English science. Of the marine monsters he said that "such are the strange combinations of form and structure in the plesiosaurus" that they seemed assembled from random parts of other creatures.[37]

In addition to finding marine reptiles, Mary Anning also found a bizarre, winged reptile. The Reverend Buckland examined and named it *Pterodactyl*. In his paper, Buckland reached into mythology to find language appropriate to describe such an unusual animal. He referred to it as "a monster resembling nothing that has ever been seen or heard of upon earth excepting the dragons of romance and heraldry."[38] The marine fossils of Lyme Regis, along with large terrestrial reptiles (eventually known as dinosaurs), being found in England in the 1820s and 1830s brought forward the notion that monsters existed in the flesh rather than just in folklore and classical literature and that science held the key to understanding them. The fossil record showed that the ancient world teemed with monsters of many different sizes, shapes, and dispositions. The discovery of the fossil remains of such animals let loose a flood of romantic theorizing about the earth's distant past, a flood that captivated naturalists and the public alike and helped renew interest in monsters.[39]

Daedalus and *Alecton*

During the 19th century, warships regularly reported encountering monstrous creatures while patrolling the world's waterways, sometimes even firing on them. Such cases added veracity to the sea-serpent question as common wisdom held that naval officers were reliable witnesses above any outrageous story telling. While several warship and sea-serpent encounters are on record, the two that stand out as especially intriguing are that of the *H.M.S. Daedalus* and the French frigate *Alecton*.[40]

In the fall of 1848, the crew of the *H.M.S. Daedalus,* under Captain Peter M'Quhae, spotted a serpent-like creature pass near their ship as they rounded the Cape of Good Hope.[41] Several officers and members of the crew saw the creature and estimated it to be 30 or 40 feet long, with a mouth full of "jagged teeth" and moving along at 15 miles an hour. This description mimicked that of the Gloucester Sea Monster. The captain and crew observed the creature but made no attempt to approach or capture it. After a few minutes paralleling the ship, the creature disappeared under the waves never to be seen again. Upon arriving back in Plymouth Captain M'Quhae gave a detailed account to *The Times* which ran the story on the ninth of October.[42] The good Captain's statements in the media drew the attention of the British Admiralty who queried M'Quhae on the incident. He replied on the 11th telling the Navy the same thing he had told the *Times*. A few days later the *Illustrated London News* ran a story with an artist's conception of the creature.[43]

The following November, the naturalist Richard Owen entered the fray with a letter to the *Times*. He argued against the serpentine nature of the creature and suggested the sailors had gotten their descriptions wrong. Having made more than his fair share of additions to the roster of living things, Owen said that "I am far from insensible to the pleasure of the discovery of a new and rare animal." The problem for him, however, was that there were only eyewitness reports, no physical remains: no bones, no skin, and no fossils to support the existence of these types of creatures. He then added a bit of a jibe that "a larger body of evidence from eyewitnesses might be got together in proof of ghosts than of the sea-serpent."[44] Captain M'Quhae was not pleased with Owen's assessment. He insisted that "I deny the existence of excitement or the possibility of optical illusions." He saw what he said he saw and was peeved that Owen would question his authority.[45]

The other naval encounter most discussed was more violent and French. On November 30, 1861, the armed French steam, dispatch boat, *Alecton* encountered a sea monster and apparently did battle with it off Tenerife in the Canary Islands.[46] The crew fired on it and managed to harpoon it.[47] After getting a rope around it the crew tried to haul it on board, but the beast's body broke up and sank and they only saved a tentacle. The *Alecton's* captain, Frédéric Bouyer, sent a report to the French consul at Tenerife which was forwarded to the French Academy of Sciences where it caused much interest and made it into the public discourse.[48] The description of the events were taken as authoritative and trust-worthy because, again, they were naval officers. The details "are statements made by men who, by their intelligence, character, and position, are entitled to respect and credence, and whose evidence would be accepted without question or hesitation in any court of law."[49]

Why this incident is important is because the *Alecton* did encounter a legendary monster, the Kraken. While it is unclear what the Daedalus encountered, it is likely the French ship came across a specimen of *Architeuthis*. Because of its deep-living lifestyle, these creatures rarely come to the surface. The few ever captured were dead individuals who floated to the surface and washed up on shore. The *Alecton* "fought" with an already dead animal. At the time, however, the *Alecton* crew thought it had battled a living and dangerous monster.[50] While a squid accounts for the *Alecton's* foe, what did the *Daedalus* and other ships see?

Richard Owen

Several high-profile naturalists commented both positively and negatively on the existence of sea monsters.[51] The most well-known and influential 19th century member of the British natural history establishment to become involved with them was Richard Owen (1804–1892). Originally studying medicine, Owen soon transferred his enthusiasm to natural history. He did extensive work cataloging existing collections of biological and fossil organisms which gave him great insight into comparative anatomy. He rose in position and influence to superintendent of the natural history collection of the British Museum, and then spearheaded the effort to make that collection its own museum. As his political connections grew, he became the tutor to the children of Queen Victoria. He coined the term *Dinosauria* to describe the large terrestrial fossil vertebrates then just being unearthed in the English countryside.[52] Due to his public profile and political connections Owen was sent materials on sea monster encounters from civilians and the government. He kept track of these and collected original letters and newspaper items for study. His collection went back to the 1830s which he kept in a special scrapbook, and in which he wrote extensive marginal notes which give useful insight into his position.[53] His notes help explain how not only Owen but British science in general began the drift away from belief in monsters as a reality. When the excitement over the *Daedalus* sighting reached the highest levels of the Royal Family, Albert, The Prince Consort, his interest piqued, asked Owen what he thought the crew had seen. The naturalist replied that he thought it likely a misidentification of a large sea lion or seal. A bit disappointed, Albert humorously dubbed Owen the "sea-serpent killer."[54]

The story gained Charles Darwin's interest as well. Still on friendly terms with Owen, Darwin read the various articles in the press. Like Owen, Darwin did not believe sea-serpents were genuine creatures, most likely sightings of conventional organisms. His sea voyaging on the *Beagle* gave him a certain familiarity with ocean life. Though he had seen many unusual and fascinating sea creatures, he never saw anything remotely like a sea monster. After reading Captain M'Quhae's report, Darwin wrote to Owen that "I never heard anything so astounding as the logbook of your H.M.S. Diddleus." [sic][55]

Along with Captain M'Quhae, others questioned Owen's analysis of the *Daedalus* creature. The physician naturalist Charles Cogswell wrote in the journal *Zoologist* that "It grows more and more necessary every day to acknowledge the *existence* [his emphasis] of a vast form of marine animal bearing some resemblance to a serpent."[56] Cogswell supported the Captain's veracity as a man of honor who would not lie or confabulate. He did question Owen's seemingly off the cuff dismissal of the creature as a sea lion. "Owen," he said, "would determine its true affinities in a moment," rather than after careful consideration and investigation.[57]

The sightings continued. In February 1849, on the heels of the *Daedalus* affair, Owen was shown the remains of another monster, this one acquired by the Duke of Northumberland. After examining it Owen declared it an example of the rare Ribbon Fish. In 1857 the British merchant ship *Castilan* reported observing a sea-serpent off St. Helens Island in the South Atlantic. Owen rejected the sighting as a misidentification. An irate correspondent to the *Cape Argus* replied that "the marine monster does exist, in the face of the deliberately recorded opinion of the greatest living zoologist Professor Owen."[58] The captain of the *Castilan*, George Henry Harrington, wanted to meet Owen to convince him of the veracity of the event. An encounter with a sea monster near Bermuda in 1860 resulted in the supposed capture of the creature, or at least parts of it. "Three separate spines of this fish were sent over to the Board of Trade, and thence to Professor Owen."[59] In an 1877 newspaper article "The Sea-serpent Again" the anonymous author said "there is not the slightest doubt that there are monsters in the sea of various kinds yet unknown to man."[60] Owen scribbled in the margins, "what proof of their monstrosity?"[61]

He followed monster developments from America as well. When the Gloucester monster reappeared off New England in 1875 – now called the Swamscott Monster – Owen wrote his thoughts in his notebook. Stranger than the notion of a monster itself, Owen wrote, "what is stranger still, not one of the country's ships that traverse all parts of the oceans from the equator to the PA [Polar Arctic] ice barrier, have ever come upon the floating carcass of a sea-serpent!"[62] He wanted to know why no hard evidence had turned up. Building upon his growing reputation as a deflater of sea monsters, the magazine *Punch* ran a satirical poem which read in part, "who killed the sea-serpent? 'I,' said Professor Owen."[63]

Owen biographer Nicolaas Rupke argues that Owen rejected the existence of sea-serpents as part of his wider plan to situate the physical evidence found in museums as centers of scientific authority. It had become common practice with sea-serpent sightings to hurriedly collect written eyewitness reports of the events preferably before a local magistrate, lawyer or other official government representative. This was thought an acceptable way of proving the veracity of a creature's existence. Reputable witnesses backed up by the imprimatur of law could not possibly be challenged. Owen felt this profoundly objectionable. He generally dismissed such "evidence" in favor of physical remains, little of which was forthcoming. In 1818, the Gloucester Monster seen off the coast of Massachusetts had been established in the eyes of some naturalists by legal documents rather than physical evidence. Naturalists in America and England accepted the creature based upon this alone.[64] Many monster incidents took this approach, but Owen wanted physical evidence confirmed by a museum-based authority not the rival expertise of jurisprudence.

Of those who disagreed with Professor Owen over the reality of sea monsters some agreed with him on other issues. By the 19th century, the debate on whether living things had been divinely created or if they had come about by some natural mechanism – often referred to as transmutation rather than evolution – grew in intensity. Sea-serpents appealed to both sides of the argument. A British sea-serpent proponent, and creationist, was Philip Henry Gosse (1810–1888). Beginning his career as a Methodist preacher, Gosse became a dissenter after embracing the notion of the imminent Second Coming. In his heart, however, Gosse was a naturalist with an intense passion for marine and other forms of biology, and a deep desire to reconcile scientific knowledge about the earth with Christian scripture. He wrote *An Introduction to Zoology* in 1843 to cater specifically to a creationist viewpoint. He later wrote a respected series of books on the birds of Jamaica and helped bring about the Victorian craze for home aquariums. This led to Gosse being made a Fellow of the Royal Society.[65]

In 1857, in order to spread his views on science and religion, Gosse published *Omphalos: an attempt to untie the geological knot.* A distinctly quirky reading of the divine origins of man as well as all life on Earth, Gosse said the fossil record did not show any evidence of transmutation. The apparent great age clearly shown by geology – and argued by Edward Hutton, Charles Lyell, and others – was all just an illusion placed there by a God who wanted the earth to look old. The title of the book comes from the Greek word for navel. To bolster his argument Gosse said that the Biblical Adam did indeed have a navel despite not being born of a human mother. God gave Adam a navel in order to show a connection between him and the human race. For Gosse, not only Adam's navel but all of creation might be an illusion and may be very ancient or just as plausibly be only a second or two old. Any age apparent in the natural world was a prop used by God. Gosse did not ridicule transmutation as many of his fellow creationists did rather he simply argued the evidence of natural history showed divine intervention. He was careful not to get caught up in the usual creationist fixation on excruciating minutia or argue specifics as to the actual age of the earth. He was content to argue the creationist point of view and bask in the notion that his audience accepted the creationist point of view as well.[66] That the book had virtually no readers or supporters did not deter Gosse from continuing his work.

In 1861, following the publication of Charles Darwin's *On the Origin of Species*, Gosse turned his attention to sea monsters. In *The Romance of Natural History* he approached a host of zoological wonders and oddities. He wrote in the light, conversational, and pithy style. The knowledge Gosse imparted was meant to educate, fascinate, and generate thoughtful reflection on man, the earth and its inhabitants, and their relationship to God. The interested investigator should go out into nature and become captivated to her thrall in the hope of learning new things. "There is so great a variety in the objects he [the naturalist] purses, and so much uncertainty in their presence," Gosse said breathlessly, "that hope is ever on the stretch."[67]

One of the topics Gosse dwells on in *The Romance of Natural History* is sea monsters. While not saying definitively that sea monsters existed, he said there really was nothing to say they did not. He felt that the large and growing number of eyewitness accounts of such creatures, like the *Daedalus* incident meant something had to be there. He also admitted that if eyewitness accounts were all that were needed, the *questiones vexatae* of sea monsters would have been cleared up long ago. Gosse's position was flexible enough that he could postulate that sea-serpents might be still living varieties of supposedly extinct species. Contrary to Richard Owen who rejected the notion that sea-serpent's might be plesiosaurs Gosse said that it was "my strong opinion, that it [the *Daedalus* creature] possesses close affinities with the fossil *Enaliosaurus*."[68]

Monsters and the Government

Along with naturalists and the public, the British government became interested in sea monsters as a possible source of commerce. In *The Octopus: the "Devil-Fish" of Fiction and of Fact* (1875) naturalist Henry Lee made straightforward descriptions of the giant octopus and cuttlefish and their possible use by industry. In one of the most mercenary discourses on monsters, Lee worked to weed out misidentifications and folklore from genuine reports for practical reasons of identifying the commercial value of such creatures. He argued that that if they did exist, the British government should exploit them for the benefit of the empire like any other raw material.[69] Lee wrote a number of pieces on sea-serpents taking the position that even the best documented cases of sightings by respected observers were simply misidentifications of either known or only slightly known, but straightforward organisms. Of such encounters, Lee said,

> I am convinced that while naturalists have been searching among the vertebrata for a solution to the problem [of sea monsters] the great unknown and therefore unrecognized, calamanes by their elongated, cylindrical bodies and peculiar mode of swimming, have played the part of the sea-serpent in many a well-authenticated incident.[70]

In a handbook written especially for the International Fisheries Exposition of 1883 he noted the case of the *H.M.S. Osbourne* of June 2, 1877 off San Vito lo Capo, Sicily. A royal yacht, the *Osbourne* was in the Mediterranean when its crew spotted a creature in the water. Most of the officers observed it for some time through telescopes and an official report was made and sent to the admiralty.[71] Here too Lee said they had seen a squid and mistaken it for a monster.

The *Osbourne* incident wound up in the hands of Richard Owen who was sent material in June of 1877, not long after the sighting occurred. Already dubious of the notion of sea-serpents as traditional unknown monsters, Owen looked at the material with a skeptical eye. Admiral Sir George Augustus Elliot (1813–1901), the commander-in-chief of Portsmouth (where the *Osbourne* was based) sent Owen copies of the statements made by the *Osbourne's* commander, H. L. Pearson, and several officers about the "unknown monster" they saw. Studying the letters, Owen wrote in the margins that "the object or phenomena may have been unknown to the observers," but they "were not necessarily caused by a 'monster.'" Unlike Lee who argued squid and octopi were the sources of sea monster sightings, Owen took a slightly different tack. Owen's interpretation of the creature was that the crew saw a cetacean. The descriptions, despite being given by reputable officers were "of no value to this naturalist." On another sighting in which witnesses said they saw a sea monster attacking a whale Owen admitted that "my skepticism on negative grounds received a severe blow after the positive statements *ex visu* . . . before the Liverpool magistrate by the respectable captain and certain of the crew." He was impressed by the veracity of the description, the trustworthiness of the witnesses, and that a court official placed an imprimatur on it. The details sounded familiar, however. Owen argued that what had been seen was a rare sighting of a pair of whales engaged in an amorous encounter. To this end he drew in his notebook several small cartoons – he labeled whale coitus – of a pair of mating whales; their flippers entwined one below the other.[72] This unusual sight was what the observers saw. Not a titanic battle between whale and monster, but cetacean procreation.[73] Taking to his notebook again, Owen asked about skeletal remains from the thousands of dead sea-serpents which must have accumulated over the years.

And what to me is strangest of all, knowing that 300 or more joints may have formed the long backbone of each of the thousands of deceased individual [sea monsters] not a single vertebra has been sent for my inspection from any shore: nor can I hear of any such specimen having been forwarded to any museum . . .[74]

As a kind of remedy to these sorts of false sightings, Owen referred to the official admiralty instructions on dealing with unusual marine life forms. The Royal Navy had long since realized the importance of collecting information on natural history from various locations around the world and had prepared a document to guide ships' officers on how to do this (Charles Darwin's famous voyage on the H.M.S. *Beagle* was an example of this philosophy). So impressed with this idea and how it could lead to better reports, Owen copied out one particular paragraph he felt was significant for clearing up the sea monster question.

When an animal is seen afloat, and is remarkable for its magnitude or other peculiarity, and is not captured, its nearest approach to the ship, its mode, course, and rate of progress, and the parts actually visible should be noted with the utmost accuracy. If the observer feels conscious that he has not the zoological knowledge requisite for determining the species from the phenomena, he should abstain from giving the animal any special name. A shot fired, if it do not hit, may so alarm the creature as to cause some sudden movement which may reveal more of its true nature.[75]

Along with the fisheries department, the Royal Navy wanted its ships to keep a keen eye out for monsters and wanted them to know what to do in the event of encountering one.

By the 1870s Owen had fallen out with many of the younger leading lights of British natural history. For various reasons – including a certain pomposity and vindictiveness on his part – people like Charles Darwin, T.H. Huxley, and Richard Dalton Hooker had all but turned their backs on the once great, now aging naturalist to the Queen.[76] He did have at least one younger, supportive acolyte who also happened to be interested in sea monsters.

Francis Buckland (1826–1880) had wide interests as a naturalist. The son of Reverend William Buckland, who had already described and named the Pterodactyl as well as the monstrous Megalosaur, Francis Buckland studied medicine at Oxford, but gave that up for natural history. Like his father, Francis Buckland engaged in the odd hobby of zoophagy: eating exotic animals not normally thought of as human fare such as rhinoceros and emu. In addition to his eating habits, "for many years," Buckland said, "I have taken the greatest interest in reports of so-called sea-snakes."[77] He was something of a protégé of Richard Owen and like Owen held little support for Darwin or natural selection. He felt mythical sea monsters were an amusing foolishness. In 1867 Buckland became inspector of British fisheries and gained access to all incoming information on sightings of sea monsters.[78] In June of 1877 Owen had been requested by Under Secretary of State, Robert Hall, to send information on the *Osbourne* incident to Buckland for examination.[79]

Buckland was a practical and energetic. His buoyant personal style made up for a good bit and extended to his writing. In *Curiosities of Natural History* (1865) he gives a fast paced, breezy and conversational narrative of his adventures with unusual zoological oddities. His approach was to explain what these giants and monsters were and to debunk myth and folk belief. He enjoyed going to the numerous displays and exhibitions of monsters then popular around London. He glides through making pithy comments and undermining belief in such creatures as mermaids, "nondescripts," and monster pigs. He was especially interested in mermaids. He examined one on display at a storefront on Regents Street. Skeptical of such things he said condescendingly of the body, "she certainly is a curious-looking thing." After describing it he continued that "to anyone with the slightest pretense to knowledge of natural history" mermaids are "a decided make-up."[80] Like a number of skeptics, Buckland argued the sea-serpents were in reality known, but misidentified organisms like the Dugong, cuttlefish, squid or whale. If they were new species, they should be made known and exploited for whatever benefit they could provide. "Thus then, we practical folks of the present day no longer make a wonder of the mermen and mermaids," he said. We should "simply harpoon them and boil them down for oil for the benefit of our patients and invalid friends and relations."[81]

By the end of the 19th century the hardheaded and practical pontificating of skeptics like Owen, Buckland and others ridiculed anyone who might think sea monsters real. A few writers still held out hope that sea monsters would be proven to be previously unrecorded animals with a folkloric past. J. G. Wood in his review of various sea monster stories argued that when looking for something to ridicule "there is always the sea-serpent to fall back upon, when a subject for wit is wanting."[82] Wood mentioned the many upstanding citizens, ship captains, Reverends, shop owners, military men, and judges who had seen the American Gloucester monster and the many learned men who had written treatises about such creatures. He felt it likely it was an unknown species of some kind though he rejected the idea they were relic Plesiosaurs. He pointed out that from an evolutionary point of view "the distance between the great saurians which are now only known by their fossil remains and those of the present day is too wide to be bridged by a survivor."[83] He thought it more likely an unknown species of whale that was on the verge of extinction which had been causing all the trouble.

Try as they might to show sea monsters did not exist, men like Owen had a tough job. Author Charles Nordoff (who also wrote on whaling and other marine subjects) said in the American *Harpers Magazine* the body of evidence from eyewitness reports was so impressive the creatures must exist. He reviewed all the big sightings and stories like the Kraken, *Daedalus*, and *Alecton* and argued that if the same amount of evidence for sea monsters existed for the moon being made of cheese, scientists would be designing giant knives to cut off chunks to bring back to earth. Nordoff finished by saying that "if I have not convinced the obstinate reader" that sea monsters were real, "I shall be disappointed."[84] While no physical evidence had been brought to light so far the admittedly extensive body of eyewitness reports proved too tempting for some academics.

The Great Sea-Serpent

The last gasp of scientific support for sea monsters came in the 1890s. Mainstream scientific – though not always evolutionary – thinking had been undermining belief in sea monsters in natural history circles, but there were still one or two naturalists not ready to give them up yet. Building upon a century of sightings and scholarly discussions, Dutch zoologist Antoon Cornelis Oudemans (1858–1943) wrote the most extensive and erudite book on the subject during this era. In *The Great Sea-serpent* (1892) Oudemans compiled an extensive and impressive compendium of sightings of strange ocean-going creatures. He eschewed the romantic literary flourishes of Humphreys and Hawkins and stuck to straight forward zoological discussion and hypotheses about the origins and nature of these creatures. He understood the resistance to belief in sea monsters but could not simply discount such a large body of eyewitness reports as others did.[85]

Oudemans complained that when some farmer or other non-scientific person found a meteorite, they would discard them for fear of ridicule from "learned men," who would undoubtedly say that the claim the object haven fallen from the heavens was a figment of their imagination. Seeing meteorites as a good analogy for sea monsters, Oudemans said that eyewitness reports of monsters were kept quiet for the same reason.

Understanding the limits of eyewitness reports while he championed them, he argued photographs of monsters would be better evidence because "this alone will convince zoologists, while their reports and penciled drawings will be received with a shrug of the shoulders."[86] Oudemans championed the notion – that Owen and others worked so hard to dispel – that eyewitness reports, even from the unlettered, held more weight than the theorizing of scientists who had not gone into the field themselves. In this way, he gave later cryptozoologists one of their more strongly held tropes: that a plethora of eyewitness accounts of some anomalous phenomenon was enough proof of its existence.

Despite his efforts, Oudemans gained little advance for his cause. In *The Case for the Sea-Serpent* (1930) R. T. Gould argued that Oudemans had not done as thorough a job with marine mysteries as he should have.[87] He said that despite the large body of eyewitness evidence Oudemans had collected his "principal defect" was that the copious amount of material made little difference. Oudemans, Gould said, "has erected a vast superstructure of theory which they [the eyewitness reports] are in no way fit for support."[88] Gould rejected sightings alone, or arguing for something based upon the veracity of witnesses, as proof. Physical evidence was crucial to any scientific enterprise and there just was not any on sea monsters. As will be shown later, Oudemans' work was resurrected in the later 20th century by the pioneer of cryptozoology, Bernard Heuvelmans.

Conclusion

In the literary Britain and Western Europe of the 19th century audiences responded to the stylized mix of science and art that a number of naturalist authors utilized concerning science and sea-serpents. Fascination with the growing knowledge of the earth's geological past focused on the strange and unusual: scenes of exploding volcanoes, wild plant life, bestial early men, and bizarre creatures locked in death struggles became a rage. Monsters naturally figured center stage in the great dramas of life being constructed in books on science as well as popular novels. Strange creatures from the past were paraded in texts, illustrations, and museum displays in ways consistent with circus sideshows and garish exhibits of freaks as well as biblical scenes in a way reminiscent of the earlier Cabinets of Curiosity. As the 19th century went by, the introduction of evolutionary thinking into this mix forced authors and researchers to choose sides: did sea monsters support or undermine evolution or creationism? Was the bizarre morphology of William Buckland's pterodactyl the result of the natural, mechanical process of biological development or the divine hand of God? These arguments were made with public participation and by using techniques of description and display not always associated with scholarly philosophy.[89]

When Richard Owen helped create the British Museum of Natural History, he and his colleagues were sure to include the best examples they could get of Mary Anning's and Thomas Hawkins' marine fossils. This was not just to allow the museum to boast of a world class study collection, but to draw in paying customers to stare and gape at these relics that, indeed, showed some monsters were real. Science was fighting for the role of authority in discussions of the natural world. Sea monsters helped do this. There were monsters, but not of the salty dog story type, or the testimony of reputable witnesses given before magistrates, but in the fossils now adorning museums, and the pronouncements of trained scientists. When it came to monsters, the public should pay attention to professional scientists not farmers or illiterate sailors.

What is important for the argument for the role of monster studies in evolution theory is that whether discounting or supporting the existence of sea monsters, the protagonists of this story made monsters and science whether by accident or design, close associates. At the same time, they found themselves increasingly rejected by mainstream science; monsters took on a subtle role in evolutionary studies. Hiding in plain sight, they influenced the language and imagery of the evolutionary debate, a debate that grew increasingly agitated as the 19th century progressed. Sea monsters had their own watery realm completely apart from the romance of humans. The monsters of the *Daedalus* and *Alecton* could hear the sailors cry, they smelled the sea and felt the sky, and let their souls and spirits fly, into the mystic.

Notes

1 An example of recent academic discussion of the nature of such creatures is C. G. M. Paxton, E. Knatterud and S. L. Hedley, "Cetaceans, Sex and Sea Serpents: An Analysis of the Egede Accounts of a 'Most Dreadful Monster' Seen Off the Coast of Greenland in 1734," *Archives of Natural History* 32, no. 1 (2005): 1–9.

2 While the terms sea-serpent and sea monster are often employed synonymously, they do indicate distinctions between creatures. The sea-serpent has characteristics of water-born snakes exhibiting no discernable limbs or fins with a relatively smooth and featureless body. Sea monsters do not have snake-like characteristics and take on a number of different body forms.

3 "On the History of the Great Sea-Serpent," *Blackwood's Edinburgh Magazine* XIII, no. III (April 1818): 32–42.

4 For an extensive discussion of the history of sea monster sightings see A. C. Oudemans, *The Great Sea-Serpent* (Leyden: E.J. Brill, 1892), and Bernard Heuvelmans, *In the Wake of the Sea-Serpents* (New York: Hill and Wang, 1968). Also see Richard B. Strothers, "Ancient Scientific Basis of the 'Great Serpent' from Historical Evidence," *Isis* 95 (2004): 220–38. Strothers makes an excellent analysis of reports and legends from antiquity and their influence on medieval authors.

5 Claus Krag, *Sverre – Norges Største Middelalderkonge* (Oslo: H. Aschehoug and Co., 2005).

6 James B. Sweeney, *A Pictorial History of Sea Monsters and Other Dangerous Marine Life* (New York: Crown Publishing, 1972).

7 C. G. M. Paxton, E. Knatterud and S. L. Hedley, "Cetaceans, Sex and Sea Serpents: An Analysis of the Egede Accounts of a 'Most Dreadful Monster' Seen Off the Coast of Greenland in 1734," *Archives of Natural History* 32, no. 1 (2005): 1–9.
8 The Linnean Society of London, *An Essay on the Credibility of the Existence of the Kraken, Sea-Serpent, and Other Sea Monsters: With Illustrations* (London: William Tegg, 1849).
9 Richard Ellis, *Monsters of the Sea* (New York: Alfred Knopf, 1994), 121.
10 Pierre Deny Denys de Montfort, *Historie Naturalle Générale et Particulière des Mollusques* (Paris, 1802).
11 A. S. Packard Jr., "Colossal Cuttlefishes," *American Naturalist* 7, no. 2 (February 1873): 87–94, and A. E. Verrill, "The Colossal Cephalopods of the North Atlantic II," *American Naturalist* 9, no. 2 (February 1875): 78–86.
12 Albert S. Gatschet, "Water Snakes of American Aborigines," *Journal of American Folklore* 12, no. 47 (October–December 1899): 255–60. Quote on page 256.
13 William Whewell (1794–1866) is often credited with the invention of this word. See Sydney Ross, "Scientist: The Story of a Word," *Annals of Science* 18, no. 2 (June 1962): 65–85.
14 Leonard Warren, *Constantine Samuel Rafinesque: A Voice in the American Wilderness* (University of Kentucky Press, 2004).
15 C. T. Ambrose, "Darwin's Historical Sketch-an American Predecessor: C.S. Rafinesque," *Archives of Natural History* 37, no. 2 (2010): 191–202.
16 Ambrose, "Darwin's Historical Sketch-an American Predecessor."
17 David Tatham, "Elihu Vedder's *Lair of the Sea-Serpent*," *American Art Journal* 12, no. 2 (Spring 1985): 33–47.
18 J. P. O'Neill. *The Great New England Sea-Serpent* (Camden, ME: Down East Books, 1999).
19 Charles E. Boewe, ed., *A C.S. Rafinesque Anthology* (Jefferson, NC: McFarland & Co., 2005), 232–40. Also see; William Dandridge Peck, "Some Observations on the Sea-Serpent," *American Academy of Arts & Sciences Memoirs* IV (1818): 86–91, and Jacob Bigelow, "Documents and Remarks Respecting the Sea-Serpent," *American Journal of Science* 2 (1820): 147–64.
20 "The New England Sea-Serpent," *Western Folklore* 7, no. 1 (January 1948): 67.
21 David Humphreys, *Letters from the Hon. David Humphreys to the Rt. Hon. Sir Joseph Banks Containing Some Account of the Serpent of the Ocean Frequently Seen in Gloucester Bay* (New York, 1817), 8.
22 Humphreys, *Letters from the Hon. David Humphreys*, 23.
23 Humphreys, *Letters from the Hon. David Humphreys*, 83.
24 Bigelow, "Documents and Remarks," 149.
25 Bigelow, "Documents and Remarks," 164. A search of the available Banks correspondence does not show any reference to Humphreys or Bigelow.
26 Leonard Warren, *Constantine Samuel Rafinesque: A Voice in the American Wilderness* (University of Kentucky Press, 2004), 164.
27 C. S. Rafinesque, "Dissertation on Water Snakes, Sea Snakes, and Sea-Serpents," *American Monthly Magazine and Critical Review* 1 (October 1817): 431–5.
28 Chandos Michael Brown, "A Natural History of the Gloucester Sea-Serpent: Knowledge, Power, and the Culture of Science in Antebellum America," *American Quarterly* 42, no. 3 (September 1990): 402–36.
29 Anonymous, "On the History of the Great Sea-Serpent," *Blackwood's Edinburgh Magazine* III (April–September 1818): 32–42.
30 Brown, "A Natural History of the Gloucester Sea-Serpent."
31 Philip Henry Gosse, *The Romance of Natural History* (Gould and Lincoln, 1861), 311.

32 See, John C. Greene, *American Science in the Age of Jefferson* (Ames: Iowa State University Press, 1984); Brooke Hindle, *The Pursuit of Science in Revolutionary America, 1735–1789* (Chapel Hill: University of North Carolina Press, 1956); and Dirk Jan Struit, *Yankee Science in the Making: Science and Engineering in New England from Colonial Times to the Civil War* (New York: Dover Publications, 1991).

33 Shelley Emling, *The Fossil Hunter: Dinosaurs, Evolution, and the Woman Whose Discoveries Changed the World* (Palgrave-Macmillan, 2009) and Thomas Goodhue, *Fossil Hunter: The Life and Times of Mary Anning, 1799–1847* (Academica Pr Llc, 2004).

34 Everard Home, "Some Account of the Fossil Remains of an Animal More Nearly Aligned to Fishes Than Any Other Classes of Animals," *Philosophical Transactions of the Royal Society* I (June 1814): 512–13.

35 Everard Home, "Reasons for Giving the Name Proteo-Saurus to the Fossil Skeleton Which Has Been Described," *Philosophical Transactions of the Royal Society* II (1815–1830): 113.

36 Henry de la Beche and William Conybeare, "Notice of the Discovery of a New Fossil Animal, Forming a Link Between the Ichthyosaurus and the Crocodile . . .," *Transactions of the Geological Society of London* 5 (1821): 559–94.

37 William Buckland, *Geology and Mineralogy Considered with Reference to Natural Theology* (London, 1836), 202–3.

38 William Buckland, "On the Discovery of a New Species of Pterodactyl in the Lias of Lyme Regis," *Transactions of the Geological Society of London*, 2nd Series, 3 (1829): 223–36. Quote on page 229.

39 On the popularity of fossil monster artwork see, Martin Rudwick, *Scenes from Deep Time: Early Pictorial Representations of the Prehistoric World* (University of Chicago Press, 1995).

40 The *Alecton* has been variously labeled a frigate, corvette, and gunboat. The famous illustration of its encounter with the monster has it as a steam, side paddle wheeler. In Greek mythology Daedalus was a skilled artisan while Alecton was one of the Furies.

41 The H.M.S. *Daedalus* was built in 1826 but was immediately put into storage. It sat idle for 18 years until it was put into service, though not before it was cut down into a smaller class of vessel. It went from a 46 gun Frigate to a 16 gun Corvette. It entered active sea duty in 1844.

42 *The Times* (October 9, 1848).

43 *Illustrated London News* (October 28, 1848): 264–6.

44 *The Times* (November 11, 1848).

45 Captain Peter M'Quhae, "Response to Professor Owen," *Zoologist* 6 (1848): 2323–4.

46 Frédéric Bouyer, "Voyage Dans la Guyane Française," in *Le Tour Du Monde* (Paris: Hachette, 1866), 273–88. Also see, *Voyage La Guyane française: notes et souvenirs d'un voyage exécuté en 1862–1863* (Paris: Hachette, 1867).

47 The captain suggested they shot at it with muskets while Willy Ley said "heavy guns." The *Alecton* did carry a pair of largeish deck guns. Willy Ley, "Scylla Was a Squid," *Natural History Magazine*, June 1941.

48 Ellis, *Monsters of the Sea*, 121. Jules Verne's *Twenty Thousand Leagues under the Seas* was published in 1869 in serial form eight years after the *Alecton* incident. As a result, I would argue it was the French ship that inspired the French author rather than the British one from twenty years before. The *Alecton* was built in 1861 and weighed 570 tons. It encountered the monster on its first major voyage. It did service around South America at French Guiana and Guadalupe until it was decommissioned in 1883, then scrapped in 1884.

49 Henry Lee, "Sea Monsters Unmasked," in *Fisheries Exhibition Literature*, vol. III (London: William Clowes & Sons, 1884), 40–1.
50 Paul Hatcher and Nick Battey, *Biological Diversity: Exploiters and Exploited* (John Wiley & Sons, 2011), 44.
51 Harriet Ritvo, "Professional Scientists and Amateur Mermaids: Beating the Bounds in Nineteenth-Century Britain," *Victorian Literature and Culture* (1991): 277–91. For a compendium of naturalist commentaries see, Bernard Heuvelmans, *In the Wake of the Sea-Serpents* (New York: Hill and Wang, 1968).
52 The standard biography of Owen is Nicolaas A. Rupke, *Richard Owen: Victorian Naturalist* (Yale University Press, 1994). Rupke also examines Owen's work on sea-serpents.
53 Richard Owen, "Notebook: Collection of Newspaper Cuttings Relating to the Alleged Appearance of the 'Sea-Serpent'," OC36, Owen Papers, Library, Natural History Museum, London. From here known as *Owen notebook*.
54 Richard Owen, *The Life of Richard Owen* (London: John Murray, 1894), 334.
55 Charles Darwin to Richard Owen. February 24, 1849, *Darwin Correspondence* (letter #1228).
56 Charles Cogswell, "The Great Sea-Serpent," *Zoologist* 6 (1848): 2316–23.
57 Cogswell, "The Great Sea-Serpent," 2320.
58 *Cape Argus* 1, no. 21 (March 14, 1857).
59 "The Great Sea Snake," *Land and Water*, September 21, 1872: 191.
60 'The Sea-Serpent Again," *Standard*, September 18, 1877.
61 *Owen Notebook*.
62 *Owen Notebook*.
63 Quoted in Owen, *The Life of Richard Owen*, 325.
64 For the New England sea monster see; J. P. O'neill, *The Great New England Sea-Serpent* (Camden, ME: Down East Books, 1999), and Chandos Michael Brown, "A Natural History of the Gloucester Sea-Serpent: Knowledge, Power, and the Culture of Science in Antebellum America," *American Quarterly* 42, no. 3 (September 1990): 402–36.
65 See Jorge Luis Borge, "The Creation and P.H. Gosse," in *Other Inquisitions, 1937–1952* (University of Texas Press, 1964), 22–5, and Ann Thwaite, *Glimpses of the Wonderful: The Life of Philip Henry Gosse* (Faber & Faber, 2002).
66 Philip Henry Gosse, *Omphalos: An Attempt to Untie the Geological Knot* (London, 1857).
67 Philip Henry Gosse, *The Romance of Natural History* (Gould and Lincoln, 1861), 271.
68 Gosse, *The Romance of Natural History*, 368.
69 Henry Lee, *The Octopus: The "Devil-Fish" of Fiction and of Fact* (London: Chapman & Hall, 1875).
70 Henry Lee, *Sea Monsters Unmasked* (London: William Clowes & Sons Limited, 1883), 7.
71 A good description of the *Osbourne* sighting can be found in *Chamber's Encyclopedia: A Dictionary of Universal Knowledge*, new ed., vol. IX (Philadelphia: J.B. Lippincot, 1901), 283.
72 In 2005 marine biologist C.G.M. Paxton and his associates reassessed the Hans Egede sighting and concluded that oft referred to event was also a case of misidentifying a whale, possibly involved in some sexual congress. See Paxton, *Cetaceans, Sex and Sea Serpents*.
73 *Owen Notebook*.
74 *Owen Notebook*.
75 Sir John F. W. Herschel, *A Manual of Scientific Inquiry: Prepared for the Use of Officers in Her Majesty's Navy* (London: John Murray, 1851), 415.

76 Nicolaas A. Rupke, *Richard Owen: Biology without Darwin*, rev. ed. (Chicago: University of Chicago Press, 2009).
77 "The Great Sea-Serpent," *Land and Water*, June 2, 1877: 196–8.
78 G. H. O. Burgess, *The Curious World of Frank Buckland* (London, 1967).
79 *Owen Notebook*.
80 Francis Buckland, *Curiosities of Natural History* (London, 1865), 134.
81 Buckland, *Curiosities of Natural History*, 145.
82 J. G. Wood, "Trial of the Sea-Serpent," *Atlantic Monthly* 53 (1884): 799–814.
83 Wood, "Trial of the Sea-Serpent," 812.
84 Charles Nordoff, "A Plea for a Monster," *Harper's New Monthly Magazine* XXI (July 1860): 178–93.
85 A. C. Oudemans, *The Great Sea-Serpent* (Leiden: E.J. Brill; London: Luzac & Co., 1892).
86 Oudemans, *Great Sea-Serpent*, preface.
87 R. T. Gould, *The Case for the Sea-Serpent* (London: Philips Allan, 1930), 5–9.
88 Gould, *The Case for the Sea-Serpent*, 8.
89 Ralph O'Connor, *The Earth on Show: Fossils and the Poetics of Popular Science, 1802–1856* (Chicago: University of Chicago Press, 2007).

5

THE TRUE BELIEVERS

A major limiting factor with monster hunters is that they never developed a methodology or philosophical approach. It is the case that monster studies did contribute to the establishment of the modern sciences of biology, zoology, and ornithology. Once those disciplines began to form, however, they lost interest in monsters. This came about because of the scientific or naturalist need for facts and evidence showed that the vast majority of monsters were the result of misidentification of known animals or the wild speculation of myths and legends. As a result, in the 21st century, there are no guidelines for how to really go about this monster hunting business. Beyond wanting to find a monster, cryptozoologists had no plan. Centuries of believers stumbling around the dark places of the human experience looking for beasts armed with nothing but semantics have found none. No Yetis, no Centaurs, no Sciapodes, and no werewolves have shuffled or slouched or stomped their way into reality. Yet, true believers have not been put off.

Monsters of the Text

Monsters appear in every culture from the classical age to the present: they likely go back even earlier than that. It stands to reason that our earliest human ancestors also feared everything around them. Unfortunately, ancient humans had no written language (that would not come around until at least 3,000 BCE and the invention of writing). They probably wanted to talk about monsters as much as we do today but needed a way to record it. The first opportunity they had, they began to write down tales of monsters. The oldest known written text, the Epic of Gilgamesh, is not only the earliest written literature but also the earliest monster story. Our ancient ancestors

DOI: 10.4324/9781003488767-6

certainly had some form of spoken communication. They undoubtably told monster stories. Unfortunately, there is scant evidence for what they spoke to each other about.

There is an intriguing example of ancient people telling stories about monsters, however. In the Cave of the Trois-Frères, Ariège, France, there is drawing of a creature that could be a spirit, an animal or a god, but that which looks like a monster. Known as The Sorcerer, it has baffled anthropologists since Henri Breuil published a drawing of it in the 1920s. Later researchers have questioned Breuil's interpretation and even his drawing, arguing that in the darkened cave he missed details and added parts that were not there. Made around 13,000 BCE, it is a complex picture part human, part animal, with a crazy bug-eyed face. Whatever the drawing is it is an indication that early humans saw scary things in the dark of night, just as we do today.[1]

As the power of modern science and historical study was brought to bear upon monstrous creatures to explain them, an interesting thing happened. This methodology only served to diminish belief by learned persons. Using a flashlight had shown the bump in the night to be a loose window shutter. Despite the turning away by scholars in the belief in the reality of werewolves. for example, true believers still argued their case. Monster enthusiasts of the late 19th and early 20th centuries didn't go out into the woods in search of monsters. Thay would have to wait until the end of the 20th century to do that. They were more like ancient authors; they searched libraries and mythology. A small coterie of English writers interested in the outré published in the late 19th and early 20th century a series of books on werewolves that had influence upon monster enthusiasm. So, I think it's going to be a long, long time.

Montague Summers

Hailing from Bristol, England, Montague Summers stands out in a field of outlandish monster hunters and enthusiasts. The 1934 review of his book *The Werewolf* in the journal *Man* said of the author that, "it is unfortunate that a writer of such erudition and of so recondite a literature should be so lacking as he is in the spirit of scientific inquiry, for he is not only uncritical but definitely obscurantist."[2] Montague Summers made a career of pretending to be one character or another, obscuring his real self. So, the vague outlines of ghosts and monsters were a natural field for him. He enjoyed, to the point of enchantment, behaving in such a way as to confound and fascinate people as to the real nature of his literary topics, but also his persona. One of the great eccentrics of English letters, Augustus Montague Summers (1880–1948), is difficult to pin down. An author of great skill and a theologian of dubious qualifications, Summers produced a body of work on the arts and history in a wide range of topics from treatises on Shakespeare to the first

English translation of the infamous witch finder's guidebook *Malleaus Malificarum*. He became a popular social animal with his caricature appearing in the London *Evening Standard*.[3] His work on monsters and other esoteric subjects, however, can be read less as an historical examination than as a performance in words.[4]

Born in Bristol, England, the second son to a wealthy father, Montague Summers grew up around books. His favorite retreat as a child was the family's comfortable and well stocked library. He spent some of his favorite childhood hours engrossed in books. As a boy he attended the local Clifton College and in October 1899 entered Trinity College, Oxford, as a Commoner.[5] Despite his famously gregarious later life as a literary gadfly, during his college days Summers was oddly absent from school life. He lived in college rooms only briefly then he took lodgings off campus. He loved Oxford, however, saying the only other places in the world to compare were Rome and Venice. He referred to Oxford as "a city of ghosts" even claiming to have seen one there.[6]

His school entry as a commoner is toward the bottom in the college registry book. This indicates a poor overall performance as a student. He exhibited that behavior, not uncommon to the creative, of doing intense scholarly work on their own, but unable to do so in a classroom. That he lived off school grounds, his name not appearing on the roles of any school clubs or activities, including the Oxford University Dramatic Society, even the lack of a school photograph all point to a detachment from mainstream college life, and a preference for other spheres.[7] Summers graduated from Oxford with a Fourth Level degree in theology in June of 1903 (In June of 1906 he was granted the MA in theology). After Oxford, Summers attended Lichfield Theological College, Staffordshire. While surviving records from this institution are sketchy, they reveal Summers entered Lichfield with a BA from Oxford in the Lent Term of 1903.

By 1908 Summers was ordained a deacon of the Anglican Church and curate of the Bristol diocese of Trinity Bitton.[8] Shortly after this he found himself immersed in a scandal involving allegations of sexual improprieties with boys.[9] His sexuality is as difficult to ascertain as any other aspect of his life. He may well have been a homosexual, or he simply may have enjoyed the ambiguity of letting others think what they wanted without confirming anything one way or another and letting the legends grow. His eccentricity extended to his person as well as his beliefs. He dressed in distinctive and quirky outfits suggestive of an 18th-century Christian cleric, yet it was of his own design, not of any recognized order. His voice was of a high pitch and pronounced with a certain laconic air and fashionable, biting wit: overall a kind of British Truman Capote. According to his biographer Father Brocard Sewell, Summers was indeed ordained a deacon by the spelunking, Anglican Bishop of Bristol, George Forrest Browne (1833–1930) in 1908. Following the supposed sex scandal, he converted to Roman Catholicism in 1911 and after some course work was ordained a Catholic Deacon. He went to Italy and found a Cardinal who ordained him a priest in an unorthodox process. Unfortunately, few of the records concerning these events are extant.[10] A historiographic issue with any study of Summers' life is the lack of original manuscript materials, correspondence and personal records, verifying any of it.

Summers began writing on occult subjects in the 1920s in the waning days of the late 19th century Occult revival. A largely British powered movement – it saw a resurgence in interest in the outré and mystical as a result of, amongst other things, dissatisfaction with traditional Christianity. The movement's central instigation came from an embrace of Eastern philosophy, magical ritual and practice, pre-Christian medieval witchcraft, and the fashionable resurrection of Egyptian and other ancient cultures grafted onto a middle-class English sensibility. Its most well-known gurus were people like Madame Blavatsky, Dion Fortune, and most infamously popular Aleister Crowley. They resurrected – or created out of whole cloth, depending upon which side one was on – religions like Theosophy, Thelema, and Wicca, and had a special love of initiatory societies which they exercised with such groups as The Hermetic Order of the Golden Dawn.

While some busied themselves with the practice of occult activities, some authors began to publish books on the history of witchcraft and spiritualism. An early and influential scholarly text on the occult came in the form of French historian Jules Michelet's *Satanism and Witchcraft* (1863). An English language translation of *La Sorcière* of the year before, Michelet (1798–1874) argued that medieval occult practices could be seen as a popular turning away from Christian orthodoxy and control and led by women. A contemporary of Summers, English archaeologist Margaret Murray became the most influential modern writer on the topic with her *The Witch Cult of Western Europe* (1921). Her work promoted the controversial idea that the murderous European witch crazes had been an attempt by the Church to crush a theological rival to Christian hegemony that could trace its origins back hundreds of years and followed carefully laid out practices. Murray (1863–1963) included detailed descriptions of the lifestyle of witches as wise women and priestesses of an ancient religious practice with its Sabbaths, covens, and horned god. Murray's work contributed much to the modern notion of the form and practice of witchcraft. Both Michelet and Murray were criticized by scholars partly for taking notions of witchcraft practice too much at face value and without enough historical skepticism. Critics faulted Murray in particular for misrepresenting and misunderstanding her sources, for painting a misleading picture of occult history, and for simply making things up.[11]

Montague Summers stands as a quirkier and more theatrical exponent of occult studies (as so many did Summers used Murray's *Witch Cult* as a source but found her analysis problematic). Writing not just as a scholar, but an entertainer, Summers harks back to the literature and theatrical works he so loved reading as a child in the family library. He wanted to impart historical information to his readers, and to convince them such things were real, but he wanted them to enjoy reading the way he did. Michelet and Murray wrote of witchcraft and esoteric arts, but not monsters – at least not in the context of this study. Summers too was interested in the more ethereal aspects of the occult, but he took on monsters in the form of vampires and werewolves.

Summers interest in writing books on witches and werewolves, officially at least, began with his relationship with editor C. K. Ogden. The editor of the hundred plus volume International Library of Psychology for the publisher Routledge, Ogden (1889–1957) was a respected literary man. A linguist and philosopher, Ogden inaugurated the *Cambridge Magazine* while still a student and established a career in English letters working with George Bernard Shaw, Siegfried Sassoon, Thomas Hardy, and other members of the English gliterati. In 1909 he founded the "Heretic's Society" at Cambridge. Ogden also edited Routledge's History of Civilization series as well as books for the Kegan Paul company, who would publish Summers' early work.[12]

Over lunch in 1923 at the Royal Societies Club in London, Ogden asked Summers to contribute a volume to the History of Civilization series. He produced *The History of Witchcraft and Demonology* (1926) and *The Geography of Witchcraft* (1927).[13] Summers wanted to clear up what he thought were the many misconceptions and misunderstandings about the occult. He delved deeply into European folklore, trial transcripts, and personal recollections. He ransacked libraries as well as histories from across time and around the world particularly the extensive collection of such works at Oxford's Bodleian Library as well as the British Library in London.[14] His bibliography for *The History of Witchcraft* ran to a thousand entries of often obscure works in many different languages. He told Ogden that "there is nothing of the kind in English."[15] He took pleasure in resurrecting forgotten lore whether that of Restoration dramas or werewolves.

By the early 1930s Summers tackled monsters. He returned to Oxford in 1929, which allowed him to work on *The Werewolf* with the collections of the university at his fingertips. He lived in a building close to Trinity at 43 Broad Street just across the street from the Old Bodleian Library (not long after he left in 1934 the block was torn down to make way for the New Bodleian library, a building Summers referred to grimly as "that superfluous monstrosity"[16]). Broad Street is also where W. B. Yeats lived. Yeats (1865–1939) the Irish author, Nobel Prize winner, and occultist could out eccentric Summers. He was a member of The Hermetic Order of the Golden Dawn and partook in the in-fighting and breakup of the society and remained a nominal member until his death. In Oxford on Broad Street Yeats held court receiving visitors from the magical and literary worlds.

The 1930s were a fascinating time for the fantastic in Oxford. Along with Yeats, in mid-decade *The Inklings*, the literary group that eventually produced *The Hobbit* (1937), *The Lion, the Witch, and the Wardrobe* (1950), and others formed. The city fairly wallowed in otherworldly story ideas and philosophies. C. S. Lewis biographer A. N. Wilson said that "silliness, sham and mumbo-jumbo have never been absent from the Oxford scene."[17] Organizationally, *The Inklings* were never more than an informal group of Oxford authors and Dons who met to get drunk, gossip, and read to each other their latest works and ideas. Central members included J.R.R. Tolkien, C.S. Lewis, Charles Williams, and a host of others. The work of *The Inklings* had a profound effect upon the literature of fantasy and the outré. They met most famously at the Eagle & Child, and other pubs around town in a loose affiliation of friends all interested in similar motifs and literary ideas. Summers was a bit early to all this, and several years older than most of *The Inklings* (Lewis was in his early thirties, Tolkien in his early forties) only Charles Williams was Summers' age. Also, Summers had left the city by 1934, so was not in the position to join the fun.[18] The only known connection between *The Inklings* and Summers is from Charles Williams (who was also a member of a Rosicrucian offshoot of the Golden Dawn). The author of occult related novels such as *Greater Trumps* (1932) about the tarot, and *Descent into Hell* (1937), Williams (1886–1945) wrote a series of non-fiction works late in life on the occult including *Witchcraft* (1941). A compilation rather than a straight forward history, Williams acknowledged using Summers' popular translation of the *Malleus Maleficarum* as a major source work.[19] Williams did not arrive in Oxford until 1939, when the Oxford University Press moved its London operations there because of the war. Other than this, Summers and *The Inklings* seemed to have never interacted despite their close physical as well as intellectual proximity. Aside from reasons already mentioned, Summers may have considered them too fantasy oriented rather than historical, and they may have considered him just too outrageous and occult obsessed. Summers seems one of those who delves deeply into fringe ideas and builds up a reputation for such but is hesitant to interact with other travelers in the same realm. What he did interact with were the libraries of Oxford.

After plowing through vast reams of printed sources at the Bodleian, and intense writing, Summers released *The Werewolf* in 1933.[20] From the beginning of his text, Summers argues that werewolves are real in the traditional sense of humans who can shape shift into wolves. He faults past authors for not taking the subject seriously enough. "They did not realize," he says, "that werewolfery was a terrible and enduring fact."[21] He comes at the topic from a philosophical and theological point of view as the only viable way. As has become so common amongst authors on the fantastic, Summers sees himself as a scholar yet denigrates academic historians and scientists who also investigate the topic. Like the theologian he pretended to be, Summers lectures the boffins for missing the point "if they disregard the science of God for the science of man." Academic anthropologists will never write well on the subject of werewolves because "they cannot read the riddle" as they are too hidebound by their educations. He himself, however, falls into the trap of arguing that universality of an idea and sheer quantity of reports give legends factual viability. Skepticism itself is a problem; the rationality of skeptical scholars keeps them from seeing the truth.[22]

Summers breaks werewolves into two sub-groups. The first is the Lycanthropes and includes those individuals who suffer from a mental disorder in which they only believe they have transformed into beasts. He dismisses them as they are deranged and not actual monsters. The other group is the "Constitutional Werewolves." They are individuals who physically transform by various occult processes and rituals – especially satanic worship – into wolves, then run off and perform various crimes against both man and God.

Having established that werewolves are real, are Satan worshippers, and are dismissed by scholars who should know better, Summers gets on with the meat of the action, the textural proof. In a way reminiscent of the likes of Edward Topsell and the Early Modern virtuosi, Summers explains wolf biology and behavior then recounts Biblical and classical sources where wolves are described as savage and bloodthirsty. He then regales the reader with a truly impressive mountain of legends and stories of werewolf incidents, both well-known and obscure. Starting with the classical world he moves through the Middle Ages and Renaissance. He apologizes for being a bit Eurocentric. He wanted to do a global treatment, but time and space would not allow. In the modern era he discusses the Beast of Gévaudan, Jean Grenier, and a case from 1925 where a local French policeman shot and killed a boy who was acting like a werewolf. He finishes with the legend of Peter Stumpp, including the first complete modern transcription of the English language pamphlet on Stumpp. He ends the book rather abruptly with a short section on werewolves and literature but drawing no conclusion other than werewolves are real.

Summers deserves credit for bringing an enormous body of work on werewolves to the public. His approach, despite his avowal of philosophy, is completely textural rather than empirical. He never quite left his father's comfortable and secure study. Summers spent hours in Duke Humfrey's reading room at the Bodleian going through piles of ancient tomes on werewolfery.[23] As with all his books, Summers indulged his love of obscure and archaic language. Pretentious words such as "mournival," "pervestigation," "expiscate," and "to farse and bombast" all appear in the first five pages of *The Werewolf*. His approach generated interesting responses.

The author of the *Man* review gave Summers credit for having compiled an impressive bibliography on lycanthropy but faulted him for making sweeping and unsubstantiated statements of "unpardonable inaccuracy." Summers argued for example that "it is recognized" that insanity is really a form of demon possession. The reviewer pointed out that Summers, despite his erudition and extensive bibliographic machinations still disregarded entire blocks of accepted modern anthropological scholarship.[24] Summers' work is oddly historiographic: he dredged up reams of ancient texts yet mostly ignored modern ones. It has become conventional to commend his scholarship but condemn his analysis.

One of the great questions about Summers' occult writing is whether he actually believed in the efficacy of what he wrote about. Historian Jacqueline Simpson's portrayal of Summers is typical of the historiography of the occult. She refers to him as "the eccentric and bigoted Catholic writer Montague Summers" who maintained that witches had actually worshipped Satan, flew on brooms and did the other nefarious activities attributed them.[25] His gullibility is so over the top it is sometimes difficult to believe he really takes it all so seriously. In 1931 Summers edited an anthology of weird tales called *The Supernatural Omnibus*.[26] In his introduction he craftily asks if an author on occult subjects need be a true believer in the subject in order to write an authentic book. Will an author, for example, write more convincing macabre tales if he is a believer in the reality of ghosts? He felt this way because "were I not myself convinced of the sensible reality of apparitions."[27] In *The Galanty Show* he states unequivocally that as far as the reality of witchcraft is concerned "there could be and there is no sort of doubt concerning" its reality.[28] Complicating this he loved to be evasive when asked about his occult leanings. He gleefully created a mystique about himself as priest and occultist. The outrageousness of his texts forces his readers to either accept or discount them almost as if he is testing his readers to see how gullible *they* are. Biographer Sewell argues "Summers was a born actor."[29] His writing can be taken as a performance on paper as much as anything done on stage. Along with his unique dress he played a role and thoroughly enjoyed himself doing it. One of his favorite roles, and the one which later devotees of the outré have gravitated toward most, is that of practitioner of the Dark Arts.

An examination of the extant materials, frustratingly limited, suggests Summers' reputation as an occultist is based on bluff and guile. Little evidence exists to show he did anything more than read and write about it. He did know Aleister Crowley, but to what extent is unclear.[30] Crowley (1875–1947) saw himself as a prophet of a new age. He established himself as a member of the esoteric Hermetic Order of the Golden Dawn, a group which included occult virtuosi such as Samuel MacGregor Mathers, A. E. Waite (who devised the modern Tarot deck) and an exotic cast of members and hangers-on. While shrouding himself in the cloak of fashionable occultism and outrageousness, Crowley took part in drug use, sexual experimentation, and the cultivation of an extravagant public persona. His life has been reevaluated in the early 21st century by scholars looking past the stereotypes to see a more complex character.[31]

In his biography of Crowley, the author and poet C.R. Cammell discusses how he personally knew both Crowley and Summers while all three lived in the affluent London suburb of Richmond.[32] Cammell was the editor of the magazine *The Connoisseur* for which Summers had written articles on rare books. Summers moved to London after leaving Oxford in 1934. Once there he altered his lifestyle. Where once "he had been a lion of the gay cultured society which gathered about [the celebrated American turned English socialite] Lady Cunard," he now turned into a semi-recluse spending his time in a stylishly appointed house with his extensive book collection.[33] Approaching old age, he reconstructed that wonderful period of his childhood and the warm and protective lair of his father's library retreating from the outside world.

Cammell's narrative is vague on the relationship between the two occult enthusiasts. He states that Crowley and Summers "had met several times in the past" but had not seen each other in the years prior to Cammell bringing them together for a sort of occult *tête-à-tête* at his apartment. In Crowley's unpublished diary there is a single entry dated July 5, 1929, London, where Crowley describes meeting Summers for the first time. The encounter went well as Crowley says, "Dinner with Montague Summers – the most amusing evening I have spent in decades!"[34] At another time, however, when asked why he and Summers did not see each other more often, the wily magician said with a hint of playful sarcasm that if they did meet Crowley would "change him into a toad."[35] Cammell also claims that Summers was deeply interested in Crowley and his work to the point of collecting newspaper clippings and following his career. Cammell remembers the two having spent the evening at his house in robust and convivial conversation talking of many mutual interests "with sparkling wit and good-fellowship."[36] Other than these few fleeting occasions, there is little evidence that Crowley and Summers did anything other than have a mutual interest and respect for one another as fellow flouters of social convention.

In the extensive extant corpus of Crowley writing and correspondence there is no mention of Summers, other than the aforementioned diary entry. Despite Cammell's insistence on Summers being an admirer of Crowley, the one time he speaks of him in *The Galanty Show* is rather minor and not especially flattering, or it may be an example of sarcasm on Summers' part for "the wickedest man in the world" as Crowley liked to be called. In 1930, Crowley now on the back end of his career, received an invitation from the Oxford Poetry Society to give a public lecture on the life of Gilles de Rais. Crowley thought the 15th century French soldier, compatriot of Joan of Arc and serial child murderer, wrongly accused. After advance advertising brought the proposed talk much attention, worried Oxford officials cancelled it. This only created more public interest. Crowley responded by publishing his lecture hoping to cash in on the titillation factor. Summers dryly commented on the anti-climax of what many expected to be a racy and controversial text. "It was bought with itching curiosity," Summers said, "and was read with disappointment" once the dry nature of it was revealed.[37] In the end, one of the major differences between the two men is that Crowley for all his mysticism looked to the future while Summers looked steadfastly to the past. Despite his eccentricity, Montague Summers made a significant contribution to the writing on werewolves. He resurrected long forgotten texts and attempted – with varying success – to make lycanthropy a scholarly pursuit.

Sabine Baring-Gould

The next in the triumvirate of English writers on the strange, Sabine Baring-Gould (1834–1924), was born in Lew Trenchard, Devon, England. Like all the authors mentioned here, he lived a varied and interesting life. From a family with deep roots in British history, unlike Montague Summers who play acted as a priest Baring-Gould was a legitimate Anglican priest being ordained in 1864. He became, with the help of his family's money and position, Curate of Horbury Bridge. He had a lifelong interest in religious music going so far as to write the wildly popular "Onward Christian Soldiers." He did not marry until he was in his early thirties then to a farm girl who worked on his property, and who was barely half his age.

Gould begins the book with a strange personal anecdote. "I shall never forget a walk I took one night in Vienne . . ." in the France countryside.[38] He had been out exploring some ancient megalithic stones near Champigny on the Rhone River and lost track of time.[39] He thought he would have to walk all the way back to his hotel. He stopped at a small local inn hoping to find a ride. The assembled townsfolk were aghast. You can't walk alone through this landscape at night, it's much too dangerous they all said. You could encounter a *loup-garou*: a werewolf. He dismisses their concerns and starts off walking. That the townsfolk were so adamant about the existence of this murderous creature he couldn't help think there might be some danger (there were probably wolves in the region but homicidal man beasts?). Just in case, he armed himself with a stout branch. He makes it to his hotel safely, but this experience caused him to want to explore the legend and produce this book.

The story unfolds just like the opening scene from *An American Werewolf in London* (1981) where the two protagonists are told by the denizens of the country pub not to go out on the moor at night. They do anyway and suffer the consequences when they are attacked by a werewolf. Following his dramatic introduction, Baring-Gould begins recounting stories from the ancient world then jumps to the Nordic tradition. Then he relates the stories of individual werewolf encounters.

One story concerns the trials of two men, Pierre and Michael, in the French countryside. They are charged, arrested, and put on trial for lycanthropy. Pierre is the first to break and tells his story about a storm scattering his flock of sheep. "There then rode up three Black horsemen . . ." One of the horsemen tells him his master will look after the sheep. Pierre has a special salve rubbed on him and this changes him into a werewolf (it seems more wolf than man). He then makes an allegiance to Satan. No mention is made of where this special salve comes from. After being coated with this salve, "I believed myself then to be transformed into a wolk."[40]

In addition to werewolves, Baring-Gould goes on at some length on the life of the infamous French nobleman Gilles de Rais (1405–1440).[41] This is the same accused child rapist and murderer that Aleister Crowley was so enamored of.[42] He was, it was claimed an alchemist and devil worshiper who tried to summon a demon. He was successful, but the demon turned down de Rais' offer of wealth and instead demanded parts of dead children. This supposedly kicked off de Rais' murder spree. Baring-Gould never comes out and says whether he believes the story or not. He included this story in a book on werewolves for a reason, or reasons he never divulges. As a wolf he attacked a little girl "and ate her up." Baring-Gould then simply ends the story with no conclusion (he does that often in this book, his narratives are rarely concluded, they just end).

He goes on to the story of the legendary young French werewolf cannibal Jean Grenier (-1574). Baring-Gould is fixated by French werewolf stories. According to Baring-Gould it is the Grenier case that drove researchers to ascribe mental aberration rather than occult forces which caused lycanthropy. Taking a modern view, Baring-Gould agreed that "medical men seem to have regarded it as a form of mental malady to be brought under their treatment rather than a crime to be punished by law." It did bother him that there were still people roaming the countryside craving human blood. He worried that, once confined to a "madhouse" they might escape and return to their bloodlust.[43]

Elliot O'Donnell

The final member of this group, and whose life mirrors Montague Summers is Elliot O'Donnell (1872–1950). He wrote ghost stories and indeed, stands as an early ghost hunter as well as an expounder on werewolves. His *Twenty Years' Experience as a Ghost Hunter* (1917) rang true in part because, he claimed, his experiences and beliefs were authentic. He believed paranormal phenomena were real.[44] Born in Bristol, England, like Montague Summers O'Donnell entered the local Clifton College as a "day boy" or non-resident student in 1884. Like Summers, he did not perform in an especially outstanding way. He did not go in for sports, rather he took to the idea of acting. Otherwise, he was described in official school documents as "work weak, rather idle" at first and "very slovenly untidy [but] takes some pains" toward the end of his time there.[45] Following Clifton College, O'Donnell entered the Queen's Service Academy in Dublin, Ireland.[46]

Like all these authors, he led an unusual life. His father, Rev. Henry O'Donnell, may have been murdered in Abyssinia while on a religious retreat. The younger O'Donnell claimed to have seen ghosts as a child and argued he was related to several famous ancient Irish Chieftains. He served in the British Army during World War One where he started getting writing work. He wrote for a number of outlets including the well-known *Weird Tales*.[47] Appearing in *Weird Tales* (11:2) February, 1928 O'Donnell's "The Ghost Table" ran alongside H. P. Lovecraft's "Call of Cthulhu," and Robert E. Howard's "The Dream Snake."

After spending some time in North America as a police officer, he began writing occult themed novels including *For Satan's Sake* (1904). He then moved to non-fiction writing on ghosts, taking traditional accounts said to be of actual spirit encounters and adding the literary and dramatic flair of a natural storyteller. These proved popular with readers and led to a career as a prototype ghost hunter and early radio commentator on paranormal subjects. O'Donnell seems to have put as much energy and creativity into his autobiography as his books on ghosts. Unfortunately, many of his personal papers vanished or were possibly destroyed intentionally by O'Donnell himself. As a result, the biographical material that is available is located in various parts of his ghost books including *Twenty Years as a Ghost Hunter*. It is here he claims extravagant descent from famous Irish warrior clans. Most scholars see him as a fabulist where his own history is concerned.

O'Donnell's foray into monster writing came in 1912 with *Werwolves*. Insisting on spelling the word with two instead of three "e"s, he explicitly believed the creatures to be genuine. "Not only were there such anomalies as werewolves formerly," he states with conviction, "but that in certain restricted areas, they are even yet to be encountered."

He then begins listing by geographic location, the many stories and reports of werewolf activity from the ancient world to the present.[48]

O'Donnell argues that the ability for an individual to transmute into a wolf – along with Second Sight, Divination, Astral Projection and other psychic abilities – was given to man by "malevolent forces" at the time of creation. He lays out a scheme in which there were several creators involved in bringing the universe into being. One God was the God of goodness and love while the other was that of darkness and evil. This malevolent creator is responsible for various paranormal abilities, including the ability to become a werewolf. This malevolent force is also responsible for evolution, which O'Donnell concedes is an operating biological force albeit a negative one. Evidence of all this, he says, is that "I have seen several sub-human and sub-animal occult phenomena in England."[49]

His interests ranged widely over paranormal topics not just werewolves. He published relentlessly. Just a brief survey shows *Some Haunted Houses* (1908), *Haunted Places in England* (1919), *Strange Sea Mysteries* (1926), and *Haunted Waters, and Trees of Ghostly Dread* (1958). His work mostly concerns ghosts with the exception of werewolves and much less monsters. It is unclear whether these three authors knew each other personally, despite their literary affinities. They did cover similar grounds. Unlike the Inklings they did not correspond or meet for drinks to discuss their work. Nonetheless, their work was read and appreciated by many budding paranormalist enthusiasts and monster lovers.

The werewolf filled a particular niche as a man-like monster. These true believing authors tried to convince others that werewolves did exist. The world wasn't having though. This was not the end of monsters, however. As belief in werewolves went into decline other creatures stepped forward to fill that role: creatures hairier and more biologically plausible. Finally, they could be used for reasons other than frightening people.

The New True Believers

As the 20th century progressed, a new generation of believers in monsters appeared. They were not interested much in the monsters of their forefathers: Blemmyae, Centaurs, and people with one giant foot did not infest their dreams. They were a new breed looking for a new cast of monstrous creatures. They were not content to read of monsters and hoary ancient tomes but swarmed out into the field to track the creatures down.

The post WWII period saw a virtual explosion of popular books, primarily paperbacks, on various areas of paranormal and fringe thinking. Ghosts, UFOs, mind powers, and especially monsters rolled off the presses. Works such as *There are Giants in the Earth* (1975) by Michael Grumley, and *Bigfoot* (1976) by B. Ann Slate and Alan Berry (and a host of others) touted the existence of man-like monsters as a reality rather than a mythology.[50] Slate and Berry argue that "paleontology has provided evidence of likely ancestors to creatures such as Bigfoot" such as Bigfoot and similar creatures in the form of the confirmed fossil ape, Gigantopithecus.[51]

Many of these books took a similar approach to the topic of monsters. They recounted an almost endless list of "true" stories about encounters with monsters. Rather than go to ancient texts, the new generation of monster enthusiasts went to what they considered even better: old newspaper reports. Scientists had faulted the modern monster hunters for ignoring primary sources. They countered by saying they have used period newspapers as their primary sources (especially American newspapers from the Old West). They completely ignored, or were unaware, of the long history of 19th century journalists making up outlandish stories to sell newspapers.[52] These new monster authors would admit that most monster sightings were misidentifications and a few outright hoaxes. In the end, however, they were left, by their own logic with the notion that there were genuine monsters out there. They argued that there was too much circumstantial evidence to dismiss monsters completely. They would often end their works with the subtle dig that if you don't believe, you are being closed minded.

Many monster hunter/researchers came from the Anglo-American world. It seems to be a Western concern to not just write about monsters but to prove them by going out and finding them. In the UK there is Nick Redfern and his work including *Man-Monkey* while America has Loren Coleman and *Bigfoot!* Coleman also opened a museum to monsters, the International Cryptozoology Museum of Bangor, Maine. One apparently universal trope for the monster hunters that seems permeate the amateur world is that mainstream scientists and scholars willfully ignore monsters. They scoffed at the stories and refused to accept anything the amateur monster hunters said and showed nothing but contempt and skepticism for the topic. The supposed rejection of monsters by scientists was, of course, not correct. In fact, some of the most well-known and respected naturalists of the 19th century, including Charles Darwin, the great evolutionist himself embraced the monstrous.

Notes

1 Peter Ucko, "Subjectivity and the Recording of Palaeolithic Cave Art," in *The Limitations of Archaeological Knowledge*, ed. T. Shay and J. Clottes (Liege: University of Liege Press, 1992), 141–80.
2 J.H.H., "Review of *The Werewolf*," *Man* 203–205 (November 1934): 182–3.
3 *Evening Standard*, ca. 1925, artwork by Matthew 'Matt' Sandsford.
4 Brian Regal, "The Occult Life of Montague Summers," *Fortean Times* 349 (February 2017): 42–6.
5 Trinity College Room Registry, and Trinity College Archive, Oxford, *School Registry Book* (n.d.), 185. A 'commoner' is an undergraduate student who attends without a scholarship.
6 Montague Summers, *The Galanty Show* (London: C. Woolf, 1980).
7 Discussion with Trinity College archivist Clare Hopkins, along with her *Trinity: 450 Years of an Oxford College Community* (Oxford: Oxford University Press, 2005), and Humphrey Carpenter, *O.U.D.S: A Centenary History of the Oxford University Dramatic Society* (Oxford: Oxford University Press, 1985).
8 Records of Litchfield College, Litchfield Records Office, admission register, D76/1/1. Thanks to Staffordshire Record Office archivist, Andrew George.
9 Robert Davies, "Summers (Augustus) Montague (1880–1948)," in *Oxford Dictionary of National Biography* (Oxford: Oxford University Press, 2004). This incident is as difficult to find source materials on as the rest of his life. It is unclear if there was a trial or just allegations.
10 *The Galanty Show*, 248. The document making Summers a deacon is in the Georgetown University collection.
11 Jacqueline Simpson, "Margaret Murray: Who Believed Her, and Why?" *Folklore* 105 (1994): 89–96.
12 J. R. L. Anderson, ed., *C.K. Ogden: A Collective Memoir* (London: Elek Publishing, 1977).
13 D. E. Wickham, ed., *Letters to an Editor: Montague Summers to C.K. Ogden* (Edinburgh: Tragara Press, 1986), 9.
14 It was digging through the British Library collection that Summers found the English language version of the original German pamphlet detailing the life of the infamous and homicidal werewolf Peter Stubb, which had been all but forgotten by then.

15 Wickham, ed., *Letters to an Editor*, 14. In the 21st century this building was rebuilt as the Weston Special Collections Library.
16 Summers, *Galanty Show*, 157.
17 A. N. Wilson, *C. S. Lewis: A Biography* (New York: W.W. Norton, 1990), 71.
18 I'd like to thank the Tolkien, Lewis, and Inklings researchers who helped me scavenge through their papers searching for any references to Summers, including Bruce Charlton, MD, of the University of Buckingham, Pieter Collier of the Tolkien Library, Belgium, Wayne Hammond, and Laura Schmidt of the Marion E. Wade Center, Wheaton College, Illinois. Their help was invaluable.
19 Gareth Knight, *The Magical World of the Inklings* (Skylight Press, 2010), 177. It is unclear what Williams was after in his membership and involvement in Rosicrucianism. Occult aficionados argue he was genuinely interested in ghosts, spirits and magic, while Christian fans prefer to think he was investigating all roads to God.
20 Montague Summers, *The Werewolf* (London: Kegan Paul, 1933).
21 Summers, *The Werewolf*, xi.
22 Summers, *The Werewolf*, xii.
23 In an homage to Summers, and because they have a collection of original works on werewolves, I too spent many long summer hours in Duke Humfrey's library researching Summers and werewolves.
24 J. H. H., "Review of *Werewolves*," *Man* 203–205 (November 1934): 182–3.
25 Simpson, "Margaret Murray," 90.
26 Montague Summers, ed., *The Supernatural Omnibus* (London: Victor Gollancz Ltd., 1974).
27 Summers, ed., *The Supernatural Omnibus*, 9.
28 Summers, *The Galanty Show*, 155.
29 Joseph Jerome, *Montague Summers: A Memoir* (London: Cecil & Amelia Woolf, 1965), 81, and Brian Regal, "The Occult Life of Montague Summers," *Fortean Times* 349 (February 2017): 42–6.
30 Lawrence Sutin, *Do What Thou Wilt: A Life of Aleister Crowley* (New York: St. Martin's Griffin, 2000), 338–43.
31 Henrik Bogdan and Martin P. Starr, eds., *Aleister Crowley and Western Esotericism* (Oxford: Oxford University Press, 2012).
32 C. R. Cammell, *Aleister Crowley: The Black Magician* (New English Library, 1969).
33 Cammell, *Aleister Crowley*, 95.
34 Thanks to Dr. Henrik Bogdan, University of Gothenburg, Sweden.
35 Sutin, *Do What Thou Wilt*, 338.
36 Sutin, *Do What Thou Wilt*, 97.
37 Summers, *The Galanty Show*, 242.
38 Sabine Baring-Gould, *The Book of Werewolves*, 1.
39 Musée et sites archéologiques de Saint-Romain-en-Gal, Département du Rhône.
40 Baring-Gould, *The Book of Werewolves*, 69–71.
41 George Bataille, *The Trial of Gilles de Rai* (Amok Books, 1991).
42 Baring-Gould uses the alternate spelling of his name as De Retz. In later years some scholars have argued that de Rai was innocent of the horrific charges. It was all a grand conspiracy to keep a good man down.
43 Elliott O'Donnell, *Werwolves* (London: Methuen & Co., 1912), 98–9.
44 Elliot O'Donnell, *Twenty Years as a Ghost Hunter* (London: Heath, Cranton Ltd., 1917).
45 Clifton College, Record Sheet. Thanks to Dr. C. S. Knighton, Principle Assistant Keeper of Archives.

46 This academy was a preparatory school for those interested in joining the police force or the military. There is an ad for the Queen's Service Academy in *The Academy* 946 (June 21, 1890): 416.
47 *Thom's Irish Who's Who: A Biographical Reference of Prominent Men and Women in Irish Life at Home and Abroad* (Alexander Thom & Co., 1923).
48 A later twentieth-century version of the true believer texts on werewolves is Marika Kriss' *Werewolves, Shapeshifters, and Skinwalkers* (Los Angeles: Sherbourne Press, Inc., 1972).
49 O'Donnell, *Werwolves*, 17.
50 See chapter seven for a further discussion of these works.
51 Slate and Berry, xiii.
52 David Dary, *Red Blood and Black Ink: Journalism in the Old West* (Knopf, 1998).

6

DARWIN AND THE MONSTERS

Monsters had been embraced by scholars for a long time by the middle of the 19th century. When Charles Darwin (1809–1882) came along, there were already questions about what they had to say about generation and deformity. He was thinking about how speciation worked and why there was so much diversity amongst living things. Monsters might have something to say about it. Not just naturalists but philosophers had monster interests. The Enlightenment philosopher Denis Diderot (1713–1784) argued, in the way that a number of classical authors did before him, that deformity cast light on non-deformity.

Diderot was less interested in monsters as biological entities as using them to counter accepted views of how the universe worked. Ever the flaunter of convention Diderot questioned accepted norms of the meaning of life. "Diderot helped transform the concept of monstrosity from metaphysical to the rational."[1] He, along with the Saint-Hilaire's and others, helped make monster studies a legitime aspect of French biological studies. Darwin did the same.

Darwin came from a long line of medical men.[2] His father, older brother and uncle were doctors. In 1825, he entered Edinburgh medical school but did not do well. He did much better at natural history courses. As a student, he collected beetles and botanized. In 1828, he transferred to Cambridge to the theology program and began to train to be a country pastor. Natural history, however, was his real love. At Cambridge, he came under the influence of several important professors who were making a name for themselves in science. The most influential of these for Darwin was John Stevens Henslow (1796–1861). It was through Henslow that Darwin ended up on the *HMS Beagle* for his momentous voyage of discovery. The *Beagle* voyage

DOI: 10.4324/9781003488767-7

was transformative for his career and his thinking about life on Earth. By the later 1830s, the idea of evolution, what he called transmutation, was taking final form in his mind.[3]

Darwin looked far and wide for inspirations and data to help prove his ideas about transmutation (the word evolution was not quite yet in wide use). He spoke with dog breeders, horse breeders, and pigeon fanciers. Anything that might let in light on biological diversity he was interested in. Darwin's work on iguanas, finches, and flowers has been studied in detail.[4] His interest in monsters is not as well known. He shared his interest in monsters with only a trusted few confidants. One of those he confided in was the British naturalist and curator of the Zoological Society of London, George Robert Waterhouse (1810–1883). They discussed whether sudden physiological changes could impact on transmutation and whether it could create a new species which is half one and half another. Darwin had read an article by entomologist G.C. Champion (1851–1927) that argued there were species that were midpoints that filled the gap between two species that exhibited characters of both simultaneously. Darwin felt this was overly simplistic. The separation between species would never be that clear cut. "I never understood . . . a halfway link he told Waterhouse, but merely one in a long series. I think you have done good service in pointing out how rare half-way-links are, if indeed they exist."[5] He also did not think that monsters were simple hybrids.

One of the issues Darwin saw with transmutation was how some species that were not very different existed in close proximity. Inspired by what he saw in the Galapagos islands, he argued the physical isolation of the place caused the finches, and other species, to slowly grow apart. Some finches, for example, developed thick beaks to crush nuts while others developed long thin beaks to reach bugs inside tree trunks. In other cases, physical isolation caused by members of a population becoming separated from the main body: by moving away or being separated by mountain ranges or water obstacles hastened changes. This environmental pressure caused speciation. But what was happening when parts of a population were not separated, yet still evolved differently? Maybe, monsters could explain it.

For Darwin and other naturalists, the term "monster" could cover a broad spectrum of cases of deformed animals and even plants. Monstrous births, sometimes called "Sports" could be profoundly deformed while others were much less so. They seemed, however, to appear suddenly with no advanced warning. Normal parents could produce Sports which might be able to bridge the gap to a new species and allow for quicker speciation. He wasn't sure, however. In his usual methodical way, he began to investigate monsters and how they appeared. In the *Origin of Species* Darwin wrote as a way of explanation that, "We have also what are called monstrosities; but they graduate into varieties. By a monstrosity I presume is meant some considerable deviation of structure in one part, either injurious to or not useful to the species, and not generally propagated."[6] Why were monstrous births isolated? Why did they not lead to many similar monsters?

He had an entire collection of sports at various museums such as the Hunterian Museum. Named for the well-known Scottish surgeon John Hunter (1728–1793), it contained hundreds of preserved specimens Darwin could look at and learn from. Darwin saw these monsters as having differences from other parts of a population that had appeared. These were not slight variations but larger differences. Monsters might provide a sudden jump to allow for speciation more quickly. As a result, Darwin had to understand the nature of these creatures. He began to share his interest in monsters with more confidants. This is when he encountered Saint-Hilaire's book.

The Saint-Hilaire's

There are few father and son research teams in the history of monster studies. But there are the father and son team of Saint-Hilaire's. The French naturalist Isidore Geoffroy Saint-Hilaire's (1805–1861) *Histoire générale et particulière des anomalies de l'organisation chez l'homme et les animaux* (1832–1837) is the work which so captivated Darwin's imagination that the usual rambunctious noise created by his children's playing and running about the house did not distract him.[7] He sat in his favorite chair reading, the daylight reflected in the glass doors of the large bookcases that covered the entire right-side wall of the study. Typical of Darwin's research style, he heavily annotated his books with a pencil: underlining passages, writing commentary and leaving notes in the margins to remind him of ideas. There were still some years to go before Darwin published *On the Origin of Species* and while he had developed the basic outline of the concept of natural selection to explain transmutation, he still worked hard at building up supportive data and continued to incorporate new ideas. As he read, he thought the Frenchman's work on mutations and birth defects might contribute to an answer as to how living things changed over time, especially over short periods of time. Saint-Hilaire attempted to explain the origins of human and animal monsters to see what could be learned about how those traits might be passed along to later generations.

Saint-Hilaire's work was one of the first modern treatises on what were then called monstrous births, now known as birth defects. A topic of fascination to learned persons and the general public alike, Saint-Hilaire wanted to take the study of monsters out of the realm of the mythological and superstitious and make it a sound science. Since the 16th century, naturalists had grasped that monsters might hold useful information on generation. To this end rather than using the term *monstrous birth* he used *teratology*.[8] At first Darwin was not sure if he should go through the trouble of getting the book. He did though and annotated his copy extensively.[9]

Isidore Geoffroy-Saint-Hilaire was the confusingly named son of naturalist Étienne Geoffroy Saint-Hilaire (1805–1861). He studied natural history and medicine. After graduation, he joined his father at the Ecole Practique giving lectures. He was elected to the French Academy of Sciences in 1833. He then held a series of academic appointments teaching about deformities, then called Monstrous Births. He rejected that term and instead began using Teratology: the study of congenital abnormalities.[10] It was at this time he began working on his multi-volume, magnum monster opus which he published as *Histoire générale et particulière des anomalies de l'organisation chez l'homme et les animaux*.[11] As a product of the Enlightenment, Etienne believed in a God, but not one who was present in his life on a daily basis. He rejected miracles and argued that monsters had completely rational and materialistic explanations: they just needed to be figured out.[12]

Saint-Hilaire argued there must be a "unity of type" for all living organisms. While Richard Owen had argued for archetypes, there were specific rules about development which did not adhere to all species. Darwin argued that "Mr. Owen suggested to me that the production of monsters . . . follow certain laws according to species."[13] Saint-Hilaire included a plethora of examples in his work on deformities of both animals and humans and attempted to organize all the varied anomalies into a coherent whole. When Saint-Hilaire discussed conjoined twins – what he called double monsters – Darwin was intrigued that "the perfect union of the two adjoining arms or legs of some of these monsters, is very striking."[14] Darwin commented that "there seems to be gradation between monsters" just as in non-monstrous species.[15] Reading Saint-Hilaire, Darwin was full of questions "how often," he asked, "have light colored monstrosities accompanied grave ones [?]"[16]

As with a number of scholars and authors before him, Saint-Hilaire wanted to make the study of monsters rational and scientific in a way it hadn't been previously. Darwin commented that "I ought to apply" Saint-Hilaire's ideas to his own work "On varieties." Unlike Richard Owen, Saint-Hilaire argued that any rules and laws of biological development that pertain to non-deformities must also apply to deformities as well. If they did not then that particular rule or law must be discarded in the study of biology. For Saint-Hilaire, it was all or nothing. Rules of biological development must apply to all organisms or none. The explanation of the imperfect must apply to the perfect as well. The author of the *British and Foreign Medical Review* states, "thus the necessary consequence of an exact and profound knowledge of anomalies will be, that the study of normal and abnormal facts, intimately associated together, will lend to each other a mutual and powerful support."[17] Saint-Hilaire did experiments where he tried to manipulate chicken eggs to artificially create mutants. He wanted to see if such mutations or monsters would influence later generations.

Darwin read widely and his use of economics, literature, animal husbandry, as well as scientific sources to support the idea of natural selection is legendary and well documented. He thought Saint-Hilaire's work on teratology might shed new light on some of the knottier questions of transmutation as well. The book and its ideas stuck with Darwin. He wrote to his friend, the botanist Joseph Hooker that, "I have just finished three huge volumes by Is. St. Hilaire on animal monsters, and a nasty curious subject it is."[18] He saw Saint-Hilaire's work as touching upon the mechanism of variation in living things – in this case humans, but which could be broadened to any organism.[19]

Part of the reason Darwin was so drawn to Saint-Hilaire's work is that the French naturalist was saying things very similar to Darwin's own ideas. Also, Saint-Hilaire seemed to take the same approach to generation as he did. In another of his marginal notes, Darwin says that for deformity such as polydactylism – which tends to show up suddenly in individuals – "Dogs & Men with polydactyle [limbs] show that germ can communicate such tendency."[20] Deformity, or Sports, could be passed along from an individual to a population. He then adds he must "must allude to this when I give my view of cause of deviations to parent treatment before impregnation."[21]

Darwin also noted from Saint-Hilaire's book the "case of a perfect [monster] individual being another head with no trunk on it . . ." He wrote in his marginal notes, "how curious this new course of the arteries."[22] The anatomy of the monster adjusted to service the multiple heads, even though the body was only geared for one head. Therefore, monster anatomy and function followed the same biological rules as non-monsters. The individual might have developed multiple heads, but the body had to adapt to it, even for a short time before the creature expired.

Darwin learned quite a bit and asked many questions about heredity and monsters. He said, "monsters occur differently in number and head [anatomy] in different species, a trunk is especially common in pigs."[23] Monstrosities, he said, are different from varieties. Monsters do not seem to come in large numbers where they are similar. The English naturalist engaged with the French author through his marginal notes in the copy he had acquired.

Darwin was not alone in his fascination with monsters, whether human, animal or plant. The scholarly engagement with monsters of all types permeates the history of biology and natural history, and contributed in subtle, but important ways to biological systemization, classification and evolutionary studies. While they did not all share the same motivations and intentions, one of the great meta-projects of natural historians all the way back to the classical world was to take nature and organize it in an understandable and rational system. Whether trying to simply arrange plants and animals on a list, to show diverse creation, or explain evolution, many natural philosophers were confronted and confounded by the need to account for monsters. Such authors as Linnaeus, Tyson, Owen, and others all studied and pondered monsters. Whether sea-serpents, sciopods, cynocephali, werewolves or Bigfoot, monsters found themselves either embraced, shoe-horned into the great chain of being or ignominiously expelled. Regardless of how they were treated, this engagement with monstrous creatures helped pave the way for the modern understanding of biology. They were not completely discarded or forgotten, however. In a fascinating turnabout, evolutionary biology influenced the mid-20th century variation on the pursuit of the monstrous, cryptozoology. Owen had direct experience with sea monsters. The problem for Owen (as discussed in Chapter 4) was that despite the eyewitness accounts he investigated for the Admiralty, there was no evidence for the existence of such creatures. The other thing that kept Owen from seeing the utility of monsters for the study of evolution was that he did no support evolution. So, he never investigated sea serpents in that way. He simply wanted to prove or disprove their existence.

The slide toward oblivion already being experienced by werewolves took on added momentum in 1859 with the publication of Charles Darwin's *On the Origin of Species*. Evolution science played a complex role in the discussion of monsters and monsters played a complex role in early discourse on evolution. Some who objected to evolution pointed to mermaids and other similar composite creatures and chuckled that belief in evolution meant belief in the possibility of such childish and superstitious phantasms. Opponents argued mermaids, the Minotaur and sea-serpents imaginary so to prove evolution, Darwin and his supporters would have to produce examples of such beasts. As they couldn't, evolution must be a sham. Anti-transmutationists argued the world should teem with such creatures if Darwin and his follower's ideas had merit. Some naturalists countered that evolution actually supported the notion of mythical creatures like satyrs and sea-serpents with a plausible scientific explanation for legends from the past. Others, unwilling to accept mermaids argued even more wondrous creatures existed as a result of evolution. *Archaeopteryx*, for example, no less fantastic than a hippogriff, existed in the fossil record. The dinosaurs and long necked plesiosaurs strode the land and swam the seas just as majestically as anything from Greek mythology.[24] Most evolutionists, however, dismissed monsters as charming creatures of folklore and superstition which Darwin helped do away with. In 1865, Francis Buckland expressed the opinion that to "anyone with the slightest pretense to knowledge of natural history [a mermaid is] a decided make-up."[25] Evolutionary biologists felt proud their work brought a deeper and more profound understanding of the workings of nature so that people would be less likely to believe in the existence of mermaids, cynocephali, werewolves and other wonders. Scientists who held evolution as an overarching paradigm rejected the notion of monsters as real because their new professionalization and reliance upon evolution theory showed these creatures did not exist, nor could they.

Charles Darwin himself makes little mention of monsters in his work. When he did, he referred to hereditary mutations in plants rather than long toothed, hairy killers. He did not address the werewolf issue directly, but circumstantially he dismissed the idea. In 1843, some years before the publication of the *Origin*, he wrote to fellow naturalist G. R. Waterhouse (1810–1888) who had just published an article on biological links. Darwin said "I never understood . . . a half way link, but merely one in a long series. I think you [Waterhouse] have done good service in pointing out how rare half-way-links are, if indeed they exist." He added, "one cannot have a simple species intermediate between two great families."[26] In general, such attitudes along with the view that dogs and primates followed two very different lines of descent proved they could in no way join together or one to turn into the other. Natural historians had also shown that while various living organisms could change their shape or color temporarily when threatening, being threatened, or locomoting, none could shape shift in the manner attributed to werewolves.

The scientific method and evolutionary theory put such a stain on monsters that by the end of the century any attempt by naturalists to engage with them in a supportive fashion could be seen as an intellectual step backward (a problem which stills haunts cryptozoology). In 1886 Cornell University physiologist Simon Henry Gage felt monsters and mythical creatures had finally been put down. "Fairies are fled," he said, "the genii banished the mermaid and the remora are captured."[27] However, this heroic narrative of science triumphing over darkness and superstition is not quite as straightforward as it appears. Science caused one monster to wither, but others to flourish.

Darwin's close friend Thomas Henry Huxley also had ideas about monsters. He Darwin told Huxley that "Monsters have frequently been described as hybrids without a tittle of evidence."[28] As both Darwin and Huxley were thinking about monsters, another English naturalist embraced them whole heartedly in a way few other had done before or since.

Thomas Hawkins

William Buckland conjured up images of mythical monsters from bygone ages while discussing pterodactyls and plesiosaurs. A protégé of his wrote what may be the most unusual book on fossil creatures and their role as monsters. Thomas Hawkins (1810–1889) was a young fossil enthusiast who wrote two books on marine reptiles which continue to resist conventional analysis to this day. *Memoirs of Ichthyosauri and Plesiosauri* (1834) and *The Book of the Great Sea Dragons* (1840) were put forward as scholarly tomes with much attention paid to morphological detail and descriptions of fossil parts, accompanied by a series of meticulously rendered drawings of the fossils, yet were written in a popular style even more poetic than David

Humphreys' letters on the Gloucester monster. Speaking of the earth's history, Hawkins waxed, "Dreaming we often crowd the events of a Century and the immensity of a Universe into the shortest space; and Ephemera born that instant to die, have most likely a sense of duration full and complete as our own."[29] His text is ethereal and romantic while it tries to be scholarly. A reviewer referred to the "exceeding bizarreries [sic]" of Hawkins' words.[30] Another reviewer said of these books, that Hawkins "may comprehend geology, we cannot comprehend him."[31] His quirky, impenetrable phraseology led the popular *New Monthly Magazine* to refer to Hawkins' text as "his occult researches."[32]

Born to a cattle dealer near Glastonbury, England, Hawkins originally studied medicine. When he was just 20 years old, his father died, leaving him a small inheritance. Like Darwin himself, Hawkins immediately gave up on medicine to pursue his true love of fossil collecting. As he lived in the vicinity of Lyme Regis he focused on marine reptiles. By all accounts, an eccentric character, he bought fossils (including some from Mary Anning) and excavated them himself as well. He managed to put together an impressive collection which drew the attention of the movers and shakers of English natural history. He came under the influence of William Buckland and became a fellow of the Geological Society of London.[33]

Hawkins' books and his attitude toward fossils pushed the boundaries of science and art as well as the boundaries between known animals and mythical monsters. His books were so expensive only a few scholars and the wealthy could buy and read them directly, while the audience level they were written at could only read of them in periodical reviews. He wanted to show himself a competent geologist and anatomist, but used words like "monster" and "dragon" in his book titles. He was not alone in this, however. Most "scientists" of the day from Buckland to Richard Owen used the same terms sprinkled throughout their work. Hawkins' subtitle, *Extinct Monsters of the Ancient Earth*, suggested simultaneously that the earth is ancient and that monsters once inhabited it. His title echoed the biblical notion that "there were giants in the earth in those days."[34] He did this in order to prove certain points.

Hawkins fell into the category, historian of science Martin Rudwick, called philosophical geologist.[35] This meant Hawkins argued how geology and fossils proved the veracity of the Bible. Like the Frenchman George Cuvier, he accepted extinction, but rejected the transmutation theory of French transmutationist, Jean Baptiste Lamarck who Hawkins referred to as "grotesque Lamark." Hawkins' mentor Buckland too saw geology as supporting scripture. In his *Reliquiae Diluvianae, or Observations on the Organic Remains contained in caves, fissures, and diluvial gravel attesting the Action of a Universal Deluge* (1823) Buckland tried to show how geological evidence proved Noah's Flood had actually happened. Likewise supporting scripture and beating back the forces of heresy Hawkins wrote "Here then, beloved Reader, we unmask an assassin [transmutation] which waylays the Doctrines of Sin, and of the Righteousness of God." These monsters, Hawkins argued, uphold the truth of biblical revelation over the error of transmutation and any other similar wrongheaded thinking.[36] Here we see the first rumblings that will in a few years explode as full on anti-evolutionism.

Another naturalist who thought geology and fossils proved the Bible was Edward Hitchcock of Massachusetts. Hitchcock (1845–1854) a geologist working out of Amherst College, Massachusetts did extensive work on the geology of New England. His most well-known work, The Geology of Massachusetts is a massive tome examining the rock formations of New England. Deeply religious despite his geological interest Hitchcock was convinced geology could help prove the Bible accurate.[37] One thing that particularly caught his eye were fossils in the vast stretches of mudflats that formed rocky layers, especially in Connecticut. New England has relatively few fossil vertebrate bones, however these once wet mudflats contained large numbers of fossil footprints. We now know these three footed prints were made by Theropod dinosaurs running across the mud. In Hitchcock's day, these creatures were unknown. Wanting to support scripture, he argued the strange bird-like prints were in fact created by the birds Noah sent out at the end of the Arks voyage to find dry land. The moderately sized prints reminded Hitchcock of bird prints. He said these were obviously made by Noah's Raven.[38] Misidentifying fossils for fanciful creatures was common.

With the marine reptiles, the fanciful language of monsters and dragons joined the technical lexicon used to describe genuine animals. Plesiosaurs and other marine life became popular characters in 19th century depictions of the ancient world which included exacting renderings of ancient fossil plant life to give the pictures depth. They were shown more often than not as mythical style monsters battling in primordial seas in epic and dramatic struggles of life and death, but as real rather than imaginary. George Richardson in *Geology for Beginners* (1843), Franz Unger in *The Primitive World* (1851), Louis Figuier in *Earth Before the Deluge* (1863), W. E. Webb in *Buffalo Land* (1872) and many others depicted marine fossils in this sensational manner.

Hawkins' eccentric enthusiasm led him to transgress boundaries in more ways than one. He wanted fossils which had aesthetic beauty as well as scientific importance. To that end, he carefully prepared his specimens by mounting them in plaster, painting them, and framing them. Missing pieces were filled in with considerable artistry. His reconstructions made it as difficult to discern where genuine and fictional material began and ended as it did in his text. Mary Anning, Gideon Mantell (who is credited with finding the first known dinosaur, the *Iguanodon*) and others worried that Hawkins was filling in details and reconstructing parts made to look like originals and was not always doing it accurately, and that this would affect their interpretation.

When Hawkins ran into financial difficulties, he sold his collection to a number of museums in the United Kingdom and the United States, including what would become the British Museum of Natural History being organized by Richard Owen. It is still not clear what parts of these specimens were reconstructed.[39]

As the 19th century progressed, the notion that sea monsters had never been anything more than stories, came under challenge. There now seemed proof, at least in the fossil record, that sea-serpents did exist in the past. A growing body of eyewitness reports suggested they still existed in the present. Sea monster lore now moved away from antiquaries and into the realm of practical men of business, science, and the military.

Conclusion

By the end of the 19th century, monsters had largely faded from the academic view. Darwin began with the idea that monsters and other deformities might help species make transitions more quickly than the slower process of natural selection. He eventually realized that deformities and "sports" impacted only on the individual not a population. Monsters might be, as he said, "a nasty curious subject" but had little impact upon transmutation. Richard Owen's work also contributed to the downfall of scientific interest in such creatures.

The Darwinian embrace of monsters, while brief, is an indication of the scientific willingness to engage with something that might have an unforeseen impact on biological diversity. Since the time of Pliny and Lucretius, monsters were not just studied but interpreted and consumed with stories told about them. As naturalists, even those wedded to what would only later be called the Scientific Method, engaged with monstrosity they did it as metaphor. As they had little else monster studies became more the study of folklore than biology. As knowledge of heredity and generation became better known and understood, the naturalist's interest in monsters waned. Science eventually narrowed its definition of "monster" and reduced it down to hereditary deformity originating in the embryo not the machinations of God, spirits or demons or some physical accident. Monster in the more visceral and traditional sense was confined to the imagination rather than reality. With the scientific rejection of monsters like the Blemmyea, Sciopods, Centars and their kin, a vacuum opened up to allow other "researchers" to pick up the pieces of monster lore. By the mid-20th century monsters were being appropriated once again for various extrascientific ends. As the new monster researchers were often not scientists or part of the intellectual mainstream, they did not feel beholden to academic thinking. They were thus free to allow their interests and imaginations to roam free over the biological landscape.

There would still be a few flutters of interest from some individual scientists, but the ship had sailed. The study of monstrous creatures moved into the laboratory. Medical researchers stopped using the term "monstrous birth" and rechristened such occurrences as birth defects. Heredity slowly became part of monster studies in an attempt to put what had been called monsters on a purely scientific basis rather than theological or folkloric grounds. However, many shapes still shambled through the forests of the world, and pulpy beasts still swam the seas. Pursuit of such organisms, both real and imagined, switched from scholars, scientists, and medical researchers to amateur investigators and adventurers.

Notes

1 G. Norman Laidlaw, "Diderot's Teratology," *Diderot Studies* 4 (1963): 105–29.
2 See Adrian Desmond and James Moore, *Darwin: Life of a Tormented Evolutionist* (Warner Books, 1991), and David Kohn, "Darwin's Ambiguity: The Secularization of Biological Meaning," *The British Journal for the History of Science* 22, no. 2 (July 1989): 215–39, and Bernard Lightman, "Darwin and the Popularization of Evolution," *Notes and Records of the Royal Society of London* 64, no. 1 (March 2010): 5–24.
3 S. M. Walters and E. A. Stow, *Darwin's Mentor: John Stevens Henslow 1796–1861* (Cambridge: Cambridge University Press, 2001).
4 Arhat Abzhanov, "Darwin's Galápagos Finches in Modern Biology," *Philosophical Transactions: Biological Sciences* 365, no. 1543 (April 12, 2010): 1001–7, and James A. Secord, "Nature's Fancy: Charles Darwin and the Breeding of Pigeons," *Isis* 2, no. 72 (June 1981): 162–86.
5 Charles Darwin to G. R. Waterhouse, December 17, 1843. Also see, G. C. Champion, "Charles Owen Waterhouse," *Entomologist's Monthly Magazine* 53 (March 1917): 67–8. This shows that Darwin unintentionally undermined the concept of werewolves by arguing you could not have a species that was half man and half dog, or any such combination such as centaurs or unicorns.
6 Charles Darwin, *On the Origin of Species* (1859): 44.
7 Isadore Geoffroy Saint-Hilaire, *Histoire générale et particulière des anomalies de l'organisation chez l'homme et les animaux*, 4 vol. (Paris: J.B. Bailliere, 1832–1837). All quotes are from Darwin's personal copy in the Cambridge University Library.
8 Girolamo Imbruglia, *Cosmopolitanism and Enlightenment* (Cambridge: Cambridge University Press, 2023).
9 Mario A. Di Gregorio and N. W. Gill, *Charles Darwin's Marginalia* (Garland Press, 1990), 306.
10 "Monsters and Teratology," *The British Medical Journal* 2, no. 1599 (August 22, 1891): 440–1.
11 The English review is in *The British and Foreign Medical Review* 8, no. 15 (July 1839).
12 Girolamo Imbruglia, *Cosmopolitanism and Enlightenment* (Cambridge: Cambridge University Press, 2023).
13 Paul H. Barrett et al., eds., *Charles Darwin's Notebooks 1836–1844* (New York: Cornell University Press, 1987), 210.
14 Darwin's marginal note in Saint-Hilaire, *Histoire générale et particulière des anomalies*, 111.
15 Saint-Hilaire, *Histoire générale et particulière des anomalies*, 308.
16 Saint-Hilaire, *Histoire générale et particulière des anomalies*, 310.
17 Saint-Hilaire, *Histoire générale et particulière des anomalies*, 310.

18 Charles Darwin to J. D. Hooker, March 31, 1845, Darwin Correspondence. Darwin's annotated copy of Saint- Hiliare is in the Cambridge University Library, *The Correspondence of Charles Darwin*, vol 2 (Cambridge: Cambridge University Press, 1986), 415.
19 Michael T. Ghiselin and Linda Jaffe, "Phylogenetic Classification in Darwin's Monograph on the Sub-Class Cirripedia," *Systematic Zoology* 22, no. 2 (June 1973): 132–40. Also see Bertrand Nouailles, "Le Monstre, la vie, l'écart. La tératologie d'Étienne et d'Isidore Geoffroy Saint-Hilaire," *Canadian Bulletin of Medical History* 35, no. 2 (2018).
20 Darwin in Saint-Hilaire, *Histoire générale et particulière des anomalies*, 500.
21 Saint-Hilaire, *Histoire générale et particulière des anomalies*, 501.
22 Saint-Hilaire, *Histoire générale et particulière des anomalies*, 246.
23 Saint-Hilaire, *Histoire générale et particulière des anomalies*, 252.
24 For an excellent study of sea-serpents in discussions of evolution see: H. Brink-Roby, "*Siren canora*: The Mermaid and the Mythical in Late Nineteenth-Century Science," *Archive of Natural History* 35 (April 2008): 1–14.
25 Francis T. Buckland, *Curiosities of Natural History* (London: Macmillan & Co., 1865), 134.
26 Charles Darwin to G. R. Waterhouse, 3 or 17 December, 1843. *Darwin On-Line*, correspondence.
27 S. H. Gage, "Zoology as a Factor in Mental Culture," *Science* 4 (1896): 209.
28 Frederick Burkhardt and James A. Secord, eds., *Correspondence of Charles Darwin* (1887): 75.
29 Thomas Hawkins, *The Book of the Great Sea-Dragons, Ichthyosauri and Plesiosauri, Gedolim Taninim, of Moses, Extinct Monsters of the Ancient Earth* (London: William Pickering, 1840).
30 Quoted in Victoria Carrol, "Beyond the Pale of Ordinary Criticism: Eccentricity and the Fossil Books of Thomas Hawkins," *Isis* 98 (2007): 225–65.
31 Percival Lord, "Review of Hawkins' *Memoirs*," *Athenaeum* 347 (1834): 469–70. Quote on page 469.
32 Quoted in Lewis Carroll, *The Hunting of the Snark: An Agony, in Eight Fits* (Macmillian, 1876), 252.
33 Arthur Bulleid, "Notes on the Life and Work of Thomas Hawkins, FGS," *Transactions of Somerset Archaeology and Natural History* 89 (1943): 59–71.
34 Genesis 6, no. 4.
35 Martin Rudwick, *Scenes from Deep Time* (University of Chicago Press, 1995).
36 Thomas Hawkins, *The Book of the Great Sea-Dragons* (London: William Pickering, 1840).
37 Edward Hitchcock, *Geology of Massachusetts* (Amherst, 1841).
38 Nancy Pick, *Curious Footprints: Professor Hitchcock's Dinosaur Tracks & Other Natural History Treasures at Amherst College* (Amherst College Press, 2006).
39 C. McGowan, "Problematic Ichthyosaurs from Southwest England: A Question of Authenticity," *Journal of Vertebrate Paleontology* 10, no. 1 (March 1990): 72–9.

7

THE INVENTION OF CRYPTOZOOLOGY

As has been shown in previous chapters, since the time of the Classical world, naturalists and other learned persons expended energy engaging with monsters. This engagement helped show the nature of birth defects, helped explain how living things procreated, contributed to an understanding of the generation of species and the relationship between humans and primates, helped expose some legends as nothing more than the creation of the human mind while others did have a biological reality. The pursuit of monstrous creatures helped create the modern fields of biology, zoology, ornithology and several others. By the beginning of the 20th century, however, scientists had squeezed most of what they could out of monsters as supposed biological realities: monstrous creatures moved into the area of philosophy and psychology as more metaphorical ideas rather than flesh and blood. By the later 20th century, monsters crawled out of sensational pamphlets and weighty texts written for high-minded scholars and onto television screens and mass media for uncritical consumption. No longer viewing the topic as a legitimate course of study for scholars, mainstream science all but abandoned it to the realm of the amateur.

This is ironic as naturalists from Lucretius to Aldrovandi and even Darwin were amateurs. The new generation did not have university training in natural history or science: in fact, they rejected mainstream science education as their forebears had embraced it. Scientific acumen and academic training now counted as a hindrance, something to be mistrusted. In addition, this new generation headed out into the field hoping to throw a net over a monster, they tracked beasts to woods much less to texts. Though they had academic university educations it was in medicine or mathematics rather than biology or zoology. It was not until the 19th century that these fields of study

DOI: 10.4324/9781003488767-8

took on formal standing. The character of the "Scientist" was being born. The term did not come into common use until after the 1830s and then only slowly.[1] The new monster scholars did not think of themselves as scientists, a term many of them would come to reject. Most preferred Naturalist, as they, unlike the scientists they claimed, were out in nature following the trails of actual monsters. Most had no university education at all, nor did they have their erudition. They were much closer, with a few notable exceptions, to being unlettered enthusiasts.

With the abandonment by the world of science, monsters lost their champions. A void appeared which was filled in short order by these enthusiastic amateur investigators who simultaneously embraced monster studies and rejected mainstream science. They were no longer interested in explaining deformities, generation, or the mechanics of evolution. Unlike Aristotle or Lucretius the new monster hunters were uninterested in attempting to work out how a given monster fit into the wider diversity of life on Earth. They treated sea monsters or the Yeti as isolated one-offs existing in a type of twilight world separate from other living beings. Like the Virtuosi of the Renaissance, they searched for mystery animals as if they were hunting trophies. With few exceptions, such as Loren Coleman's *International Cryptozoology Museum* in Portland, Maine, the new generation differed from the Renaissance Virtuosi in that they were not out to create Wunderkammers as a place of pride. You would not find special display rooms at the homes of modern monster hunters where they could show off their collections: no stuffed alligators or Jersey Devils hanging from the ceiling. The 20th century monster hunters were not really collectors. They did not scour the media for mentions of a farmer or police officer shooting a chupacabras for the purpose of then rushing to the location to try and buy the carcass for pickling and a jar. Another major difference is that the Renaissance Virtuosi enthusiastically embraced science as a worldview they wanted to be part of. Cryptozoologists in large part rejected mainstream science as some type of intellectual trap.

In the past creatures such as Scipodii, Centaurs, and sea-monsters occupied monster enthusiasts, now another cast of characters came to the forefront of interest. Tales of far-off exotic places unapproachable to Western investigators gave way to locations that could be visited. Monsters were now romantic beings, something that could be looked for outside of texts and legends. Increased knowledge of and the ability to travel to strange lands, at least strange to the predominantly White Western investigators, opened new fields and objects of study. The growing number of amateurs interested in monster studies helped create the concept of the search for unknown animals (those not formally acknowledged by science): it was the invention of Cryptozoology.

One of the first 20th century writers to come close to introducing the search for mystery animals was Willy Ley in his 1949 article for *Mechanix*

Illustrated, "Do Prehistoric Monsters Still Exist?" The German born paleontologist turned rocket expert wrote a series of books on mysterious animals. Ley used the term "Romantic" or "Exotic Zoology" to describe his work, terms which did not catch on. What was needed however was a catchier term.[2]

While dominated by amateur investigators, the formal notion of Cryptozoology was begun by two academically trained scientists in the 1950s. Belgian biologist Bernard Heuvelmans (1916–2001) began publishing on the new field with his *Sur la Piste des Bêtes Ignores* (1955) and then *In the Wake of the Sea Serpents* (1968). Ivan Sanderson (1911–1973) was a Scottish, Cambridge trained zoologist, well-known author and early television nature show presenter.[3] His *Abominable Snowmen: Legend Come to Life* (1961) became the bible of early anomalous primate studies.[4] He argued that these creatures were populations of Neanderthals who somehow survived into the modern era. Seeing each other's work they realized they were kindred spirits. Heuvelmans championed the work of A. C. Oudemans on sea monsters seeing a useful methodology and built upon his work.[5] Out of this came the term cryptozoology. Heuvelmans attributed the term to Sanderson (although it had once appeared in 1959 in the work of French author Lucien Blancou). As trained scientists, unlike their intellectual children, they both felt monsters could open avenues of research into evolution and turned their careers over to it. The bulk of amateur investigators eschewed theory and metaphor. They did not seek higher philosophical meaning. They did not scour libraries and archives for strange creatures to puzzle over. Unlike any of the investigators of monsters noted so far, this new generation of hunters wanted to go out into the wild, into the forest primeval: over the river and through the woods to capture a real-live monster. This, they thought, would be a great way to show up the scientists who in equal parts ridiculed and ignored them. What many of the amateurs didn't realize was that there were still scientists on the hunt. Maybe not as many as in the past, but still a few.

The Loch Ness Monster

Though the term cryptozoology came about in the post WWII years, the modern mania for monsters began in the 20th century in the post-WWI years in Scotland (with the Loch Ness Monster) and Central Asia (with the Yeti). It is also at this point that the interest in such creatures leaves the realm of local legend and enters the worldwide consciousness. The aquatic creature today known as the Loch Ness Monster traces its mythical lineage to the 7th century. An early and typically unconfirmed report has the creature attacking swimmers on the great lake of Scotland. The Irish Abbot Columba (521–597 CE) heard of such stories while he preached Christianity to the Picts of the region. When these attacks seemed to be increasing, Columba, eventually Saint Columba, went to the lake in 565 CE, said prayers, and with great flourish commanded the creature to stop attacking people. This seems to have done the trick as, while sightings continued, there were no more stories of monster attacks.[6]

Despite a growing number of supposed sightings, the story remained little more than a quaint, local tradition until the 1930s when an article appeared in the *Courier* about a sighting by a man named George Spicer. This brought wider attention to the legend, which now grew proportionately as more and more people became enthralled by the idea of a large, anomalous marine animal inhabiting the loch. By 1933, more people claimed to have seen the creature. Authors Daniel Loxton and Donald R. Prothero argue that the 1933 explosion of sightings coincided with the release of the film *King Kong* that year. There is a scene in the film where an aquatic creature very much like the traditional descriptions of St. Columba's beast is prominently displayed. The next year saw the publication of the so-called Surgeon's Photo. The legend now ran off and became a cryptozoological icon.[7] The genie had been loosed from the lake, never to be put back.

In 1934, gynecologist Dr. Robert Kenneth Wilson snapped what is the most famous photograph of a monster ever taken. It was published in the *Daily Mail* on April 21 of that year. It shows a sinuous, long-necked creature gliding across the lake, the bulk of its obviously large body just below the waterline. The photo caused a sensation and for years was argued over by skeptics and believers. It later turned out to be a hoax concocted by adventurer and filmmaker Marmaduke Wetherell (1883–1939) who, in an effort to hide his wheeze, asked the good doctor to pose as the one who snapped the picture. It was put together as revenge by Wetherell against the *Daily Mail* for them discounting photos of supposed Nessie (as the creature was now known) footprints Wetherell had brought to the newspaper the year before. In 1994, it was revealed that the pranksters had taken a wind-up toy submarine, covered it in wood putty to create the monster neck and head, then placed it in the loch to be photographed.

In the 1970s, intrigued by the legend, investigator Robert Rines (1922–2009) and his associate Peter Scott brought modern technology to bear upon the mystery. They employed sonar and underwater photographic equipment. Until this point all Loch Ness Monster photos – and in fact all photos of marine and lake monsters such as Cadborosaurus and Champ – had been taken above the water line. Now able to probe the murky depths of the lake, they produced the "Gargoyle Head" and "Flipper" photos. These were the most sensational Loch Ness pictures since the Surgeon's Photo and purported to show the creature close-up. Like the Surgeon's picture, the Gargoyle Head and Flipper images underwent extensive scrutiny. The photos seemed so convincing that Rines proposed giving the creature the scientific name of *Nessiteras rhombopteryx*. Investigator John Lawton pointed out with a hint of a smile that *Nessiteras rhombopteryx* is an anagram of "Monster hoax by Sir Peter s."[8]

The Snow Man

While the creature of Loch Ness is the most widely recognized aquatic monster, the most widely known types of modern terrestrial monsters are the anomalous primates. These creatures include Sasquatch, the Skunk Ape, and a host of others. The first widely known such creature was the Nepalese Yeti: a large, hairy, imposing bipedal creature that seemed to lurk and skulk and peep about the frosty regions at the top of the world. The groundwork for the Yeti legend was laid in the 1820s when journalist Henry Newman of the *Calcutta Statesman* heard about local people seeing a strange creature tramping around the snow in the shadow of Mount Everest. Sensing an opportunity, Newman wrote an article about the sightings. Newman had some trouble with the Sherpa word *Meh-Tay* for the creature. This is often translated as "That thing over there." By the time his story hit the headlines the name had been hopelessly mangled into Abominable Snowman.

The article brought out other sightings and more interest. The avalanche began when a startling photograph of supposed Yeti footprints surfaced in 1951. Taken by well-known British mountaineer Eric Shipton (1907–1977), it showed a line of strange, vaguely primate footprints in the snow.[9] News of the Shipton photo spread rapidly outside the mountaineer community. Through the 1950s and early 1960s, a series of "expeditions" to Nepal headed off to catch a Snowman. This early period was dominated by large scale operations in the same vein as traditional Western explorations to far-off places with men tramping through the snow. Sponsored as publicity stunts by organizations such as *The Daily Mail* and *Life Magazine*, they were meant to capture public interest more than anything else. In later years, this model was seen as an inefficient waste of time and resources and fell out of favor. These expeditions, however, involved a number of scientists who either wanted to go along or give advice. These included Dillon Ripley of the Smithsonian Institution, Indian ornithologist Biswamoy Biswas, and Harvard anthropologist, and racial speculator Carlton Coon. The idea of a Yeti set the scientific community thinking about the evolutionary history of primates. Though they had very little to go on, from a theoretical point of view, the Yeti must be some type of primate. If that were the case biologists and zoologists could latch onto that thread. Carlton Coon, for example, felt knowledge of the Yeti could explain some things about human evolution and behavior.[10] He kept up with Yeti developments. Amateur investigators such as the Irish tea grower Peter Byrne and the excentric Texas oil man Tom Slick, also took part. Because of some of their backgrounds included working for the OSS during World War II, and the Chinese crackdown on Tibet and Nepal going on at the time, this led to speculation as to a murky Cold War CIA involvement in the search. Despite the hullabaloo and the headlines, what little evidence these expeditions found, including the theft of the infamous "Yeti Hand," led nowhere. When it was all said and done, and the expeditions were disbanded, the Yeti remained as elusive, mysterious, and uncaptured as ever.[11]

Russian Homins

Another reserve of anomalous primates was Russia. During the Soviet Era, the government thought monster hunting was a good thing. They created a Snowman Commission to sponsor a hunt for what in Russia and Mongolia were called Almasti. One of the leading lights of Russian cryptozoology was Boris Porshnev (1905–1972). As an historian of French revolutionary history, Porshnev developed what came to be called the Porshnev Theory. This was an idea that held that creatures such as the Almasti, Yeti and others were relic populations. They were, Porshnev and others argued, small groups of a species that had moved to remote environments and remained largely unchanged. While the majority of a species either went extinct or evolved into other forms, a small group became isolated and survived as the evolutionary forces that acted on the majority of the species had much less impact, so they survived and adapted to their region.

Boris Porshnev saw the Almasti as not just creatures who could make human evolution clearer but also the progress of Communism as a better political system than Capitalism. These supposed human ancestors, when left to their own devices and allowed to progress at their own speed, would naturally follow a more inclusive and equal socialist route of development rather than an exclusive, competitive capitalist way. Monsters were not just mystery animals but had as much to say about human politics as it did biology.

The Russians, unlike Western enthusiasts, refused to call these creatures monsters. They preferred the term Homin as they saw them as closely related to modern humans. To explain the Almasti was to explain human evolution. As a result, the authorities in Russia took greater interest in the topic than did those of their Western counterparts. Boris Porshnev managed to put together an expedition to the Pamir mountains (where reports of Almasti were most common) with government backing and funding to search for the creatures in 1958.[12] They also created a special Snowman Commission through the auspices of the Soviet Academy of Science to oversee research into the topic of Homins. Unfortunately, the Pamir expedition found nothing. The Soviet government soon tired of the whole affair and the Snowman Commission was disbanded. The only tangible result was that Boris Porshnev's career was almost ruined.

The North American Giant

Just as the period of Yeti expeditions was winding down, news of a similar creature began to appear. This time, rather than the blinding snows of

Mount Everest, this creature's playland was the thick pine forests of the Pacific Northwest of Western Canada and the United States. Of all the organisms cryptozoologists searched for, the manlike monsters strike the deepest chord. Alternately known as ABSMs (Abominable Snowmen), mystery apes, and anomalous primates, the search for these creatures is as much a search for our ancient monster selves. They are the ones that draw the most attention from both academics and amateurs alike. Humans have a long history of seeing frightening creatures – so like us, yet so different – in the far-off inaccessible corners of the world, it is a fear that transcends geography, time and ethnicity. We seem to need to find out what they are.

Reported to be roughly eight feet tall and weighing 500lbs or more, Bigfoot is reported to be a hairy manlike, bipedal creature of mostly nocturnal and solitary habits. The creature is notoriously hard to find and has appeared in a small number of photographs all of which are contentious as to their authenticity. The most commonly found evidence for sasquatch and like forms are the large footprints they sometimes leave behind. Researchers have collected an enormous number of plaster casts of such prints and use them to argue for the creature's existence.

There are reports from Native American legends of Bigfoot-like creatures that preceded the arrival of European settlers. There are also scattered reports from explorers and trappers in the 18th and 19th century who encountered creatures that fit the sasquatch mold. These reports, and the idea of sasquatch, remained little known outside of folklorists and local inhabitants until the middle of the 20th century. In 1958, a construction gang pushing a road through the Six Rivers National Forest, Northern California, near a stream called Bluff Creek encountered unusually large footprints in the dirt near their heavy equipment. While they never saw who or what left the tracks, the workers claimed the tracks reappeared often. A road gang member, Jerry Crew, made a plaster cast of one of the tracks and brought it to local *Humboldt Times* reporter Andrew Genzoli. Crew told the reporter about what had happened and that the construction workers had taken to calling the mysterious visitor, Bigfoot. Genzoli put the story on the AP wire service which caused an instant sensation. The name Bigfoot carried a great pop culture punch. A number of amateur naturalists already interested in man-like monster sightings began to investigate the Bigfoot incident. Men like Canadian journalist John Green, Swiss-Canadian Rene Dahinden, the Irish born Peter Byrne (who had investigated the Yeti) and many others researched the phenomena, talked to eyewitnesses and collected evidence about these creatures. With the Bigfoot story, the floodgates opened and many people in the United States and Canada began to come forward telling how they too had seen the creatures. Along with the amateur naturalists, a small coterie of professional scientists also joined the search for these elusive animals. These included British primatologist John Napier (1917–1987), whose *Bigfoot: The*

Yeti and Sasquatch in Myth an Reality (1973) was the first major scholarly work to address the evidence that came out of Asia.

Then came the most well-known image of Bigfoot. The 16mm film footage commonly called the Patterson-Gimlin Film became, and continued to be, the marker by which all other imagery of the creature is judged. Shot in October 1967 in the Bluff Creek region of Northern California by a pair of cowboy/ranchers, Roger Patterson and Robert Gimlin, as they explored for the creature. It is in the trees, it's coming. Barely a minute in length, the PG Film, despite its jumpy nature and scratchy quality is widely held up as definitive proof of the creature's existence.

American paleoanthropologist Grover Krantz (1932–2002), after seeing the PG Film, became convinced of its authenticity. A specialist in human and primate locomotion at Washington State University, Krantz became the scientist most associated with the study of anomalous primates, and champion of their existence. For his efforts, he was roundly denounced by fellow scientists who thought the evidence was lacking and monsters were a waste of time. Surprisingly, Krantz was also ridiculed by amateurs who thought as a professional scientist he was infringing on their territory. Krantz studied the way the creature in the film walked and concluded that this was not something that could be faked by a hoaxer. He also later claimed that he identified dermal ridges in plaster casts from other sites. Dermal ridges, fingerprints on feet, were another thing a hoaxer could never fake, or even think to fake.

Krantz spent most of his 30-plus-year career working to articulate an acceptable theoretical paradigm which linked Bigfoot to the extinct Asian primate *Gigantopithecus*. Known from fossil teeth and jaw parts, *Gigantopithecus* was the largest primate known. The available fossil evidence suggests to some, though not all scientists, that *Gigantopithecus* was a large biped which weighed in the 500lbs range and stood about eight feet tall. This description was virtually indistinguishable from reports of sightings of Bigfoot. Krantz put together technical data on Bigfoot from such sightings and from the Patterson-Gimlin Film to produce a biomechanical description of the creature that made sense anatomically. He attempted to use the techniques and methodologies of physical anthropology to explain what the creature looked like and more importantly explain why it could exist. One of the key pieces of evidence Krantz relied on were the "Cripplefoot" tracks found near the towns of Bossburg and Colville in Washington State, in November 1969. These prints intrigued Krantz because of anatomical anomalies in the right foot that he felt could not have been created by hoaxers. This approach became a common one amongst enthusiasts: if the anatomical details of tracks could not be faked, some of the tracks must be genuine, thus the creatures must exist. In more recent years, anthropologist Jeffrey Meldrum of Idaho State University has taken up Krantz's mantle of the most visible professional scientist to argue for Bigfoot's existence in North America.[13]

Meldrum tried a variation on the expedition model of monster hunting: he assembled a team of scientists to look at evidence in the laboratory.

Many in mainstream biology and zoology feel that the evidence for such creatures is slim and circumstantial at best. They argue that such an animal runs counter to evolution theory and that the attempt to link Bigfoot to *Gigantopithecus* is dubious as the most recent fossils are 300,000 years old. In addition, no *Gigantopithecus* fossils have ever been found in the Americas. Another problem with the field is the proclivity for hoaxing that goes on. Following his death, Ray Wallace (1918–2001) the foreman of Jerry Crew's construction gang, was accused of having faked the famous Bigfoot tracks. His family stated that he had been faking tracks most of his adult life. Other researchers who were familiar with Wallace accepted that he was a hoaxer but argued that he had not faked the tracks found by Jerry Crew.

Bigfoot of the Text

The practice of early historians and naturalists building their work on a collection of prior authorities, many of which were classical Greeks and Romans, went back to the scholastics of the Middles Ages. Scholars were expected to make reference to Plato, Aristotle, Pliny and others for answers to various questions about the natural world or the past. Reliance on such authorities proved problematic and a growing number of intellectuals rebelled in favor of knowledge and data they had collected themselves. This concept helped bring on the Scientific Revolution and the Enlightenment. The American naturalists of the post-Civil War period had no ancient sources they recognized connected to the history of North America, certainly not to any monsters. In addition, what seemed to be an almost built in aversion to authority whether political, ecclesiastic, or scientific tended to keep the American amateur naturalists closer to empiricism than theory and past authors. That began to change when the monster hunters realized they might have as much luck searching for monsters in the garden of the text as well as forests and fields.

With the end of the long 19th century, American monster enthusiasts had little to investigate other than the New England sea monsters. By the post WWII period, fledgling monster investigators began to ransack, not classical texts, but more modern newspapers. Responding to academic historians call for primary documents, the amateurs of the 20th century went to news accounts of encounters with strange creatures. Focusing on 19th century accounts the new monster enthusiasts found many tales of human encounters with Bigfoot-like creatures, especially in the Western Territories of the United States. They were able to compile a large number of such reports including ones that seemed to suggest that the Biblically based giants, the Nephilim, had once roamed the wastes of North America. They then threw these down at the feet of academics saying, "You wanted primary sources? Well, here they are!" The problem was that the new amateurs didn't really understand American history, and especially American journalistic history. American newspaper writers from Benjamin Franklin, to Mark Twain on up to the 20th century regularly faked all sorts of newspaper stories, especially about monsters, because they knew such fanciful tales were popular and would boost sales.[14]

Following Bernard Huevelmans' lead some cryptozoology enthusiasts noticed familiar things lurking in ancient sources like Aristotle or Pliny – a few even read Marco Polo or John de Mandeville in modern English translations. Those who took this more egghead approach saw references to hairy men everywhere. Native American folklore showed itself something more than just quaint stories. They scoured old newspaper accounts and traveler's tales for references to manlike monsters in North American then world history. Bigfoot hunters ransacked American and Canadian newspapers for any reference to apes, wild men or strange human-like creatures and took heart at finding numerous historical references to them. It was one thing to have a passage from Pliny, Marco Polo or Tibetan manuscripts which referred to Bigfoot-like creatures, but another to have such descriptions turn up in old British Columbia, California, Texas or New York newspapers. These texts became the Aristotle that sasquatch enthusiasts embraced almost as gospel truth. They held these as evidence of their position arguing that sasquatch had been known well before its appearance in popular culture in the 1950s. Indeed, the majority of books on cryptozoology, starting with Bernard Heuvelmans and Ivan Sanderson, focused on collections of past reports of hairy bipedal ape-like creatures as the core of their argument. The more such historical reports they found the more they became convinced the creature real. Unfettered by the risk of professional ridicule, the amateurs continued to work combining eyewitness reports, historical references, and the myriad of tracks into what they considered physical proof that they were right and the professionals wrong.

The problem was that few of these amateurs were trained in the ways of professional scholarship to analyze these texts. They assumed that if a newspaper account from 1870, for example, said the bones of a giant were found in a cave in Utah or Nevada, or that some prospectors had an encounter with Sasquatch-like creatures, the report could be taken at face value. This was enough. They never questioned the veracity of the account. This highlights the inherent flaw of amateur based cryptozoology in the 21st century. Part of the academic training received by scientists, archaeologists, or historians is how to examine evidence, how to separate out the logical from the illogical. How to trace sources, and how to question the data. The amateurs had not been trained to question, or how to question, the evidence they encountered.

As a result, when academics were skeptical of the evidence the amateurs brought them, they didn't see it as the normal checks and balances of scholarly inquiry, they saw it as a personal attack upon them. They saw grand conspiracies. They saw it as an attempt by an entrenched hierarchy to squash new ideas and conceal important truths from the world. Professional scientists became the enemy. What most of the amateurs didn't realize was that many scholars were open to evidence of monsters if only the amateurs brought them some. There was one moment, however, where a number of professional scientists thought they had finally found a real monster.

The Beast of Minnesota

The field of cryptozoology studies is studded with frauds. Stories of encounters and sightings, faked footprints, photos, and film footage, along with claimed physical evidence such as hair samples and DNA, have been a constant since the field's earliest days. This proliferation of nonsense is the primary reason for the falloff of interest from the scientific community that began in the 20th century. Data would come in of the existence of some cryptid, the scientific community would enthusiastically investigate, then the hoax would be uncovered. One of the last major such hoaxes was the Minnesota Iceman of 1968. It marks a high point of interest from the scientific community, but like many high points it also marks a moment of steep decline. Carnival exhibitor Frank Hanson was showing a body of a hairy manlike creature in a block of ice which he claimed was a "missing link" he had acquired from an anonymous millionaire. Hearing of the creature, cryptozoologists Bernard Heuvelmans and Ivan Sanderson went to see the exhibit. They managed to get permission to examine the thing close up. They took photos and measurements, though a careful examination was impossible due to the fact the creature, or whatever it was, was frozen in a block of ice inside a glass case. They did their best and in the end, despite the difficulties and roadblocks Hansen put up, became convinced of its authenticity.[15] This began a flurry of interest from the scientific community. It was not a footprint, or a blurry photo, but an actual body: something the scientific community had been clamoring for all along.

Both Heuvelmans and Sanderson published on the find – including photos – which drew the attention of the Smithsonian Institute in the form of John Napier and later the FBI. The law enforcement agency became interested when it was thought that the creature had been dispatched by being shot with a firearm. The FBI's interest was short lived. Hanson became worried by all the official interest including a request by the Smithsonian to purchase or at least be able to examine the prize and so he took the exhibit down for some time. When it reappeared, Hanson claimed he had been exhibiting a fake all along. The creature and Hanson soon slipped into the obscurity of cryptozoology folklore when it became clear the whole thing had been a hoax that had spun wildly out of control. In the 21st century the body, or what was purported to be the original "body" went up for sale. It was purchased in

2013 and exhibited in Texas. There was no more interest in it from the scientific community. The only visitors to the sad creature now are cryptozoology enthusiasts and the curious.

A more recent hoax was the Bigfoot in a box, put forward by a pair of Georgia men in August of 2008. They put up photos of a carcass they claimed they had found while out hunting. They had scooped up the remains and put it into a large box. Much anticipation was generated, and the discoverers claimed they had sent some of the remains out for DNA analysis and that they would be holding a major press conference to announce their find. Some eagle-eyed web browsers saw from the photos that the "carcass" was actually an off-the-shelf gorilla costume with some possum entrails strewn over it. The fraud quickly fell apart, and the promised press conference came to nothing.

Other Monsters

In the 21st century, mainstream science had learned its lesson about monsters. Breathless reports of giant snakes in Africa, bloodsuckers in Puerto Rico, Mongolian Death Worms and other such things have not created much scientific interest in the way they might have in the past. In the 1990s, reports of a creature which attacked livestock and drained their blood caused media attention, but little else. The Chupacabras was said to attack barnyard animals and like a quadrupedal Dracula drank their blood. Fevered tales of the creature skulking in animal pens and having their way dropped off. A few photos of mangy dogs and coyotes were easily dismissed as such. Popobawa was a supposed spirit reported in Tanzania in the mid-1990s.[16] It is claimed that the creature attacked people and even killed some. Swahili for Batwing the creature was seen in Zanzibar as well. As with many such legends there are only stories and unverified reports, with no physical evidence. Similar to the Chupacabras, reports of Popobawa seem to come in waves interspaced with periods of no sightings.

A few scientists intrigued by monster stories tried more up-to-date approaches to monsters. New York University anthropologist Dr. Todd Disotell took a genetic approach by studying the DNA of various claimed biological remains of anomalous primates. Dr. Charles Paxton of the University of St. Andrews looked at sea monster reports as a way of studying human perception. Prolific English author Richard Freeman has traveled numerous times to Sumatra in search of the elusive Orang Pendek.[17] A diminutive anomalous primate, the Orang Pendek is thought to leave all the same sorts of evidence, hair samples and footprints, as their larger Asian and North American relations.

Not all Cryptids are primates or sea serpents. In Africa there is a supposed living dinosaur that dwells in the Congo River basin called Mokele-Mbembe. Beginning as a Bantu legend, by the late 19th and early 20th centuries it went from being a spirit creature to being a flesh and blood animal. It is mostly described as looking like a sauropod dinosaur. Cryptozoologists argue that it is a population of "relic" creatures: leftovers from the age of dinosaurs that have gone unrecorded due to the isolated habitat they live in. It has become of interest not only to cryptozoologists but to Young Earth Creationists. Christian fundamentalists have been trying, in their awkward, unsuccessful way, to prove Mokele-Mbembe real because they think if they do, it will somehow prove evolution isn't a real process. Young Earth Creationism (YEC) holds that the Earth is only 6–10 thousand years old. If there is a genuine dinosaur relic in Central Africa, that means humans and dinosaurs co-existed thus breaking down the edifice of evolution.[18] Some creationists have attempted to bring the entire idea of cryptozoology under the sway of Christian fundamentalism with varying levels of success, though not outside the Christian community. The name Mokele-Mbembe may be a result of the German influence on the history of the nation of Congo. Possibly the earliest mention of dinosaurs in Africa came in 1909. Congo was part of first the German empire then the Belgian empire. In 1978, it became Zaire. In 1925, Dr. George Sarton, the influential editor of the history of science journal *Isis*, related a story, with his tongue planted firmly in his cheek, that during World War I authorities in the Congo required any dinosaurs there to be equipped with running lights for safety.[19] University of Chicago biologist Roy P. MacKal became so fascinated by the Mokele-Mbembe legend he went on two expeditions to the Congo to try to prove its existence in 1980 and 1981.[20] He collected a number of stories from local people, and poked around the Congo River, yet found nothing.

Upon his return to the United States, MacKal joined forces with cryptozoologist Richard Greenwell (1925–2013) and Bernard Huevelmans to create the International Society for Cryptozoology in 1982. The society drew many of the bright scientific lights of the movement including Grover Krantz. To go with the society, they created a scholarly journal as a place to publish research in the field.

With a few exceptions, the monsters of Africa were pursued by Euro-American investigators rather than local, homegrown cryptozoologists. As it did when it began, cryptozoology is the realm of mostly Euro-American amateur investigators. Even the Russian end of the search has fallen into obscurity. Those professional scholars who take an interest in monsters now are mostly historians, anthropologists, and archaeologists. They, however, do not have any interest in proving these legends real or not, but examining them for what myth and stories tell us about the human condition.

So, What Is Cryptozoology?

Over the years, efforts to define cryptozoology remained somewhat vague and unformed. Bernard Heuvelmans and Ivan Sanderson never really progressed beyond, "it's the search for unknown animals." The attempt to explain cryptozoology continued well into the 21st century when the most successful articulation of the field came from Chad Arment. He called cryptozoology a "targeted-research methodology for zoological discovery." Cryptids, he said, are "ethnoknowns." These are species unrecognized or officially described by western science but are known to local cultures. Ethnoknown species "are those for which some prior contact with man has been communicated to others." It is a species' place in mythology and cultural legend that distinguishes a cryptid from other newly discovered species.[21]

Cryptozoologists are primarily focused on physically large organisms. There are no tiny cryptids. Unknown species which become accepted by mainstream science are out there. Unknown plant, insect, and aquatic species are discovered on a regular basis, but have no historic of folklorish existence, and rarely get noticed by the media.

One of the great icons of cryptozoology is the Coelacanth. A primitive Lobe Finned fish the Coelacanth is held up as the shining example of a creature thought to be extinct which was found alive. This genus was known only from fossil versions that dated to the Late Cretaceous Age and the end of the dinosaurs. Its discovery as a living organism seemed to support the "relic" concept which was thought to support the existence of everything for the Yeti to the Almasti. Unfortunately, it is a poor example. It was not known in a mythical sense. No one was looking for it until it was seen that fishermen started pulling them up in December 1938 off the coast of South Africa.[22] The population is thought to be relatively small. They are now under siege and close to man-made extinction because in the past when fishermen caught them they were thrown back as they were not eaten by humans (they tended to cause diarrhea) and had no commercial value. Then it was realized that scientists would pay money for specimens to display and study.

Despite the false starts, hoaxes and general lack of enthusiasm from professional scientists, monster researchers continue to go into the woods to find evidence of their quarry. Since the 1950s sightings have continued and have come from all over the world: Florida to New York to Oklahoma and Texas, Puerto Rico, the Congo, the British Isles to Southeast Asia and everywhere in between. Now armed with high-tech gear like night vision scopes, heat sensors, and motion triggered cameras, the monster hunters still head out into the woods like bird watchers. If creatures like Bigfoot, the Chupacabras, or monster snakes or spiders are ever found it will not be by professional scientists, but by amateur enthusiasts. Cryptozoologists pursued their quarry in the murky depths of Loch Ness, the frozen wastes of the Himalayas, the dark waters of the Congo River basin with its screeching, exotic birds; and many other places around the world. As of the early 21st century, all this effort has produced some tantalizing evidence, but no proof of the existence of these monsters.

Conclusion

The 20th century was a heady time for monster studies. A number of mystery creatures seemed on the cusp of leaving the doldrums of cultural obscurity and stepping into the limelight of scientific acceptance. The Loch Ness Monster, the Yeti, Sasquatch, and others teetered on the edge. Cryptozoologists held their breath as if watching a golf putt that had the ball rolling around the edge of the cup about to drop in. Then in a moment of heartbreak, the ball spun away from the cup to roll off into failure. So close to success, but ending up in distant uselessness. These monsters, after a tantalizing moment in the sun, retreated back into the shadows to regain their lostness.

The 20th century scientific interest in monsters is fascinating. This intense rush of excitement over monsters by scientists was not the blooming of a new era, but a last gasp. Too many professional researchers had taken the chance and publicly showed support for various mystery creatures only to be burned by hoaxes and charlatans and misidentifications. For a brief shining moment it appeared as if cryptozoology would become legitimate. No scientific insights were forthcoming, however. Its ranks, once filled with scientists, now comprised mostly true amateurs running around the woods with little or no scientific or historical training, wearing night vision goggles and looking for large hairy shapes in the dark spooked by every twig snap.

Notes

1 Sydney Ross, "Scientist: The Story of a Word," *Annals of Science* 18, no. 2 (1962): 65–85.
2 Willy Ley, "Do Prehistoric Monsters Still Exist?" *Mechanix Illustrated* (February 1949): 80–144, and *Exotic Zoology* (New York: Viking Press, 1955). Ley (1906–1969) was a rocket pioneer in the 1920s and 30s and was an associate of Werner Von Braun. With the rise of the Nazis Ley, unlike Von Braun, refused to add his expertise to the German rocket program. He emigrated to the US and became a popular science author.
3 Bernard Heuvelmans, *Sur la Piste des Bêtes Ignores* (Plon, 1955), and *In the Wake of the Sea Serpents* (Hill and Wang, 1968). Also see, Jean-Jacques Barloy, *Bernard Heuvalmans: Un Rebelle de la Science* (Les Éditions De L'oeil Du Sphinx, 2007).
4 Ivan Sanderson, *Abominable Snowmen: Legend Come to Life* (Chilton, 1961).
5 Heuvelmans helped resurrect Oudemans' now virtually forgotten work as a sort of unsung founding father of cryptozoology. A. C. Oudemans, *The Great Sea-Serpent* (Leiden: E.J. Brill; London: Luzac & Co., 1892).
6 Katja Jylkka, "Witness the Plesiosaurus: Geological Traces of the Loch Ness Monster," *Configurations* 2018, no. 2 (2018): 207–24, and Tim Dinsdale, *The Loch Ness Monster* (Boston: Routledge and Kegan Paul, 1976).
7 Daniel Loxton and Donald R. Prothero, *Abominable Science: Origins of the Yeti, Nessie, and other Famous Cryptids* (Columbia University Press, 2015).
8 John Lawton, "Nessiteras rhombopteryx," *OIKOS* 77, no. 3 (1996): 377–80.
9 Soule Gardner, "The World's Most Mysterious Footprints," *Popular Science* (December 1952): 133.

10 Carlton Coon, *The Story of Man: From the First Human to Primitive Culture and Beyond* (New York: Alfred Knopf, 1954).
11 Catherine Nixon Cooke, *In Search of Tom Slick* (Texas University A&M University Press, 2020), and Loren Coleman, *Tom Slick: True Life Encounters with Cryptozoology* (Craven Street Books, 2002).
12 Boris Porshnev, *The Soviet Sasquatch* (UK: CFZ Press, 2021). This book has a tortured publication history. Originally written in Russian in 1963. It was not widely available, but only talked about. A number of other scientists in Russia disagreed deeply with Porshnev's conclusions. His career was stifled in much the same way Grover Krantz's career was belittled by American academics. It was not until the 21st century that a copy was obtained in England, translated, and finally available more widely.
13 Jeff Meldrum, *Sasquatch: Legend Meets Science* (Forge Trade, 2009).
14 Gregory Evans Dowd, *Groundless: Rumors, Legends and Hoaxes on the Early America Frontier* (Johns Hopkins University Press, 2015).
15 Bernard Heuvelmans, *The Strange Saga of the Minnesota Iceman* (San Antonion: Anomalist Books, 2016), and Ivan Sanderson, "Preliminary Description of the External Morphology of What Appeared to be the Fresh Corpse of a Hitherto Unknown Form of Living Hominid," *Genus* 25 (1969): 259–78.
16 Benjamin Radford, "Popobawa! In Search of Zanzibar's Bat-Winged Terror," *Fortean Times* 241 (November 2008): 34–9, and Joe Nickell, "The Skeptic-raping Demon of Zanzibar," *Skeptical Briefs* (December 1995): 7.
17 Richard Freeman et al., *Orang Pendek: Sumatra's Forgotten Ape* (CFZ Press, 2011).
18 Edward Guimont, "Hunting Dinosaurs in Central Africa," *Contingent Magazine*, March 18, 2019, and Brian Dunning, "Hunting the Mokele-Mbembe," *Skeptoid Podcast* #727, May 12, 2020.
19 George Sarton to Willey Ley, May 3, 1951, box 1, Folder 5, National Space Administration Museum Archive.
20 Roy P. MacKal, *A Living Dinosaur?: In Search of Mokele-Mbembe* (New York: E. J. Brill, 1987).
21 Chad Arment, *Cryptozoology: Science and Speculation* (Landisville, PA: Coachwhip Publications, 2004), 9.
22 J. L. B. Smith, *Old Fourlegs: The Story of the Coelacanth* (Longmans & Green, 1956).

CONCLUSION

The story of monsters is a long fairytale full of unnerving eyes, sharp spines and teeth. They inhabit dark snowy forests littered with pine needles and pinecones, mountain lakes, stifling swamps, and vast primordial seas. They are observed by soldiers in medieval armor peering down from castle keeps, nervous sailors loading deck guns and unnerved when their cannon balls bounce harmlessly off rubbery, pulpy bodies seemingly unaffected by their fire. They make roaring sounds echoing across the Pacific Northwest, French forests, the mountains of China, and the jungles and waterways of central Africa. The stories have been told around roaring fires and glittering computer screens.

Charles Darwin had worked on his book, *On the Origin of Species*, since he returned from his voyage on the *HMS Beagle* in 1836. He agonized over it for years. His worry about how it would be received haunted him like Frankenstein's hideous creation haunted the Swiss scientist who created him. Both Victor Frankenstein and Charles Darwin created their monsters largely in secret. Darwin knew that once he unleashed it on the world his creation's behavior and the world's response to it would be forever out of his control. Having finally reached the point where he decided to publish his research, he wrapped his manuscript in brown paper and addressed it to his London publisher ready for the post. His research into monsters and his exposure to them helped him work out some of the mechanics of natural selection. As with the experience of biological diversity, he knew that monsters had played a powerful but subtle role in how evolution worked.

Throughout its long history, monster studies contributed to knowledge about the biological world. The great disappointment of modern cryptozoology is that, as of this writing, it has not contributed anything meaningful to science. The modern science of biology, human development, and evolution

DOI: 10.4324/9781003488767-9

walked metaphorically through the forest of knowledge. The cries of unknown birds singing and the roars of unseen animals created strange dissonant songs. Mysterious creatures lurking just beyond the blackness of the thickly growing pines and oppressive darkened swamps filled the air and the imagination. Just like the forests and foreboding places people walked through fearful of finding our worst selves, science walks through that forest all the while catching little glimpses of monstrous shapes hiding just skulking in the shadows. Always peering in, not always acknowledged, but still there just out of frame.

In the 21st century, the arguments over evolution still rage. Religious fundamentalists have gone through great lengths to show what they consider the hollow and unsupported evidence for evolution. They argue evolution makes no sense philosophically or theologically. While monsters such as Bigfoot and the Loch Ness Monster play ridiculous roles in modern anti-evolution arguments, other more ancient monsters do not. Fundamentalists, and it seems, cryptozoologists as well, never mention all those creatures once the staples of monster theory. The Sciapodes, Blemmyea, and Centaurs are the ancestors of the Yeti, Chupacabras, and werewolves, yet they seem to have been lost to the modern monster hunter vocabulary.

It's interesting to speculate what the likes of Pliny, Lucretius, or Aldrovandi would think of cryptozoology. From the philosophers of the ancient world to the scientists of the modern world, to the simply curious, they all search for monsters for the same reason. Whether Richard Owen, Grover Krantz or Charles Darwin, we look for monsters to find ourselves.

BIBLIOGRAPHY

Ambrose, C. T. "Darwin's Historical Sketch-An American Predecessor: C.S. Rafinesque." *Archives of Natural History* 37, no. 2 (2010): 191–202.

Anderson, J. R. L., ed. *C.K. Ogden: A Collective Memoir*. London: Elek Publishing, 1977.

Anonymous. "On the History of the Great Sea-Serpent." *Blackwood's Edinburgh Magazine* III (April–September 1818): 32–42.

Arment, Chad. *Cryptozoology: Science and Speculation*. Landisville, PA: Coachwhip Publications, 2004.

Ashton, John, trans. *The Voiage and Travayle of Sir John Maundeville Knight*. London: Pickering and Chatto, 1887.

Asma, Steven. *On Monsters: An Unnatural History of Our Worst Fears*. New York: Oxford University Press, 2009.

Augustine of Hippo, Henry Bettenson, trans. *City of God*. Penguin Classics, 2004: Book VIII.

Barloy, Jean-Jacques. *Bernard Heuvalmans: Un Rebelle de la Science*. Les Éditions De L'oeil Du Sphinx, 2007.

Bataille, George. *The Trial of Gilles de Rai*. Amok Books, 1991.

Bigelow, Jacob. "Documents and Remarks Respecting the Sea-Serpent." *American Journal of Science* 2 (1820): 147–64.

Birrell, Anne, trans. *Classic of Mountain and Sea*. Penguin Classics, 2000.

Bodin, Jean. *De la lycanthropie, transformation, et extase des sociers, auec la refutation des argumens que Bodin allegue au 6. chap du second livre de sa Demonomanie, pour soustenir la realité de ceste pretenduë transformation d'hommes en bestes*. Par, 1615.

Boewe, Charles E., ed. *A C.S. Rafinesque Anthology*. Jefferson, NC: McFarland & Co., 2005, 232–40.

Bogdan, Henrik, and Martin P. Starr, eds. *Aleister Crowley and Western Esotericism*. Oxford: Oxford University Press, 2012.

Borge, Jorge Luis. "The Creation and P.H. Gosse." In *Other Inquisitions, 1937–1952*, 22–5. University of Texas Press, 1964.

Bouyer, Frédéric. "Voyage Dans la Guyane Française." In *Le Tour Du Monde*, 273–88. Paris: Hachette, 1866.

Boyd, Maggie. "From the Archive." *Bulletin of the American Academy of Arts and Sciences* 76, no. 2 (Winter 2023).
Brinkman, Paul. "The Second Jurassic Dinosaur Rush and the Dawn of Dinomania." *Endeavour* 34, no. 3 (2010): 104–11.
Brown, Chandos Michael. "A Natural History of the Gloucester Sea-Serpent: Knowledge, Power, and the Culture of Science in Antebellum America." *American Quarterly* 42, no. 3 (September 1990): 402–36.
Buckland, Francis T. *Curiosities of Natural History*. London, 1865.
Buckland, William. "On the Discovery of a New Species of Pterodactyl in the Lias of Lyme Regis." *Transactions of the Geological Society of London*, 2nd Series, 3 (1829): 223–36.
Buckland, William. *Geology and Mineralogy Considered with Reference to Natural Theology*. London, 1836, 202–3.
Bulleid, Arthur. "Notes on the Life and Work of Thomas Hawkins, FGS." *Transactions of Somerset Archaeology and Natural History* 89 (1943): 59–71.
Burgess, G. H. O. *The Curious World of Frank Buckland*. London, 1967.
Cameron, Malcolm H. V. "Seventeenth Century Science with Particular Reference to Comparative Anatomy." *Canadian Journal of Comparative Medicine* 11, no. 5 (May 1947): 132–7.
Cammell, C. R. *Aleister Crowley: The Black Magician*. New English Library, 1969.
Carpenter, Humphrey. *O.U.D.S: A Centenary History of the Oxford University Dramatic Society*. Oxford: Oxford University Press, 1985.
Carrol, Victoria. "Beyond the Pale of Ordinary Criticism: Eccentricity and the Fossil Books of Thomas Hawkins." *Isis* 98 (2007): 225–65.
Cartledge, Paul, ed., trans. *Herodotus: The Histories*. Penguin Classics, 2014.
Chamber's Encyclopedia: A Dictionary of Universal Knowledge. New ed., vol. IX. Philadelphia: J.B. Lippincot, 1901.
Clifton, James. "Kunst- und Naturalien-Kammern of seventeenth-Century Naples: Johann Daniel Major's view from Germany." In *The Science of Naples*. UCL Press, 2024.
Cogswell, Charles. "The Great Sea-Serpent." *Zoologist* 6 (1848): 2316–23.
Cohen, Jeffrey Jerome. *Monster Theory: Reading Culture*. University of Minnesota Press, 1996.
Coleman, Loren. *Tom Slick: True Life Encounters with Cryptozoology*. Craven Street Books, 2002.
Coleman, Loren. *Bigfoot*! Paraview Books, 2003.
Cooke, Catherine Nixon. *In Search of Tom Slick*. Texas University A&M University Press, 2020.
Coon, Carlton. *The Story of Man: From the First Human to Primitive Culture and Beyond*. New York: Alfred Knopf, 1954.
Dandridge Peck, William. "Some Observations on the Sea-Serpent." *American Academy of Arts & Sciences Memoirs* IV (1818): 86–91.
David, Dary. *Red Blood and Black Ink: Journalism in the Old West*. Knopf, 1998.
Davies, Robert. "Summers (Augustus) Montague (1880–1948)." In *Oxford Dictionary of National Biography*. Oxford: Oxford University Press, 2004.
De Blécourt, Willem. "A Journey to Hell: Reconsidering the Livonian 'Werewolf'." *Magic, Ritual, and Witchcraft* 2, no. 1 (Summer 2007): 49–67.
De la Beche, Henry, and William Conybeare. "Notice of the Discovery of a New Fossil Animal, Forming a Link Between the Ichthyosaurus and the Crocodile . . ." *Transactions of the Geological Society of London* 5 (1821): 559–94.
De L'Ancre, Pierre. *Tableu de l'Incostance de Mauvais Anges et Demons*. Paris, 1612.
Della Porta, Giambattista. *Natural Magic*. London, 1658.
Del Lucchese, Filippo. *Monstrosity and Philosophy: Radical Otherness in Greek and Latin Culture*. Edinburgh Scholarship Online, 2019, 93–129.

Dendle, Peter, and Asa Simon Mittman. *The Ashgate Research Companion to Monsters and the Monstrous*. Routledge, 2012.

Denys de Montfort, Pierre. *Historie Naturalle Générale et Particulière des Mollusques*. Paris, 1802.

De Woskin, Kenneth J., and James Irving Crump, trans. *In Search of the Supernatural*. Stanford University Press, 1996.

Dinsdale, Tim. *The Loch Ness Monster*. Boston: Routledge and Kegan Paul, 1976.

Dudock, S. A. C. *Nicolaeus Tulp: The Life and Work of an Amsterdam Physician and Magistrate in the 17th Century*. Six Art Promotion Bv., 1998.

Dunning, Brian. "Hunting the Mokele-Mbembe." *Skeptoid Podcast* #727, May 12, 2020.

Eisler, Robert. *Man Into Wolf – An Anthropological Interpretation of Sadism, Masochism, and Lycanthropy*. Routledge and Paul, 1948.

Ellis, Richard. *Monsters of the Sea*. New York: Alfred Knopf, 1994.

Emling, Shelley. *The Fossil Hunter: Dinosaurs, Evolution, and the Woman Whose Discoveries Changed the World*. Palgrave-Macmillan, 2009.

Espinasse, Margaret. *Robert Hooke*. London: William Heinemann, 1956.

Evans, R. J. W. *Rudolf II and His World: A Case Study in Intellectual History 1576–1612*. Thames & Hudson, 1997.

Evans Dowd, Gregory. *Groundless: Rumors, Legends and Hoaxes on the Early America Frontier*. Johns Hopkins University Press, 2015.

Falcon, Andrea, and David Lefebvre. *Aristotle, the Generation of Animals: A Critical Guide*. Cambridge: Cambridge University Press, 2018, 207 and 427.

Felton, Debbie, ed. *The Oxford Handbook of Monsters in Classical Myth*. Oxford: Oxford University Press, 2024, 230–42.

Findlen, Paula. *Possessing Nature: Museums, Collecting, and Scientific Culture in Early Modern Italy*. University of California Press, 1994.

Findlan, Paula. *Athanasius Kircher: The Last Man Who Knew Everything*. New York: Routledge, 2004.

Fiona, Melhuish. *Ole Worm's Museum Wormianum*. University of Reading Special Collections, May 12, 2020.

Freeman, Richard, et al. *Orang Pendek: Sumatra's Forgotten Ape*. CFZ Press, 2011.

Friedman, John Block. *The Monstrous Races in Medieval Art and Thought*. Syracuse University Press, 2000.

Frost, Brian J. *The Essential Guide to Werewolf Literature*. Madison: University of Wisconsin Press, 2003.

Galambosova, Caroline. "What Is a Wunderkammer?" *Daily Art Magazine*, September 23, 2024.

Gardner, Soule. "The World's Most Mysterious Footprints." *Popular Science* (December 1952): 133.

Gatschet, Albert S. "Water Snakes of American Aborigines." *Journal of American Folklore* 12, no. 47 (October–December 1899): 255–60.

Giacobini, Giacomo, y Riccarda Giraudi. "Y el hombre encontró al simio." *Elementos* 68, no. 14 (Octubre–Noviembre, 2007, Página 3). In his *History of Anthropology*, vol. 1. Routledge, 1996.

Glassie, John. *A Man of Misconceptions: The Life of an Eccentric in an Age of Change*. Riverhead Hardcover, 2012.

Goldwyn, Robert M. "Nicolaas Tulp (1593–1674)." *Medical History* 5, no. 3 (1961): 270–6.

Goodhue, Thomas. *Fossil Hunter: The Life and Times of Mary Anning, 1799–1847*. Academica Pr Llc, 2004.

Gosse, Philip Henry. *Omphalos: An Attempt to Untie the Geological Knot*. London, 1857.

Gosse, Philip Henry. *The Romance of Natural History*. Gould and Lincoln, 1861, 311.
Gould, R. T. *The Case for the Sea-Serpent*. London: Philips Allan, 1930.
"The Great Sea-Serpent." *Land and Water*, June 2, 1877, 196–8.
"The Great Sea Snake." *Land and Water*, September 21, 1872, 191.
Greene, John C. *American Science in the Age of Jefferson*. Ames: Iowa State University Press, 1984.
Gross, Charles G. *A Hole in the Head: More Tales in the History of Neuroscience*. MIT Press, 2009.
Guimont, Edward. "Hunting Dinosaurs in Central Africa." *Contingent Magazine*, March 18, 2019.
Hatcher, Paul, and Nick Battey. *Biological Diversity: Exploiters and Exploited*. John Wiley & Sons, 2011, 44.
Hawkins, Thomas. *The Book of the Great Sea-Dragons, Ichthyosauri and Plesiosauri, Gedolim Taninim, of Moses, Extinct Monsters of the Ancient Earth*. London: William Pickering, 1840.
Henderson, Felicity. "Unpublished Material from the Memorandum Book of Robert Hooke, Guildhall Library MS1758." *Notes and Records of the Royal Society* 61 (2007): 129–75.
Herschel, Sir John F. W. *A Manual of Scientific Inquiry: Prepared for the Use of Officers in Her Majesty's Navy*. London: John Murray, 1851.
Heuvelmans, Bernard. *Sur la Piste des Bêtes Ignores*. Plon, 1955.
Heuvelmans, Bernard. *In the Wake of the Sea-Serpents*. New York: Hill and Wang, 1968.
Heuvelmans, Bernard. *The Strange Saga of the Minnesota Iceman*. San Antonion: Anomalist Books, 2016.
Hindle, Brooke. *The Pursuit of Science in Revolutionary America, 1735–1789*. Chapel Hill: University of North Carolina Press, 1956.
Home, Everard. "Some Account of the Fossil Remains of an Animal More Nearly Aligned to Fishes Than Any other Classes of Animals." *Philosophical Transactions of the Royal Society* I (June 1814): 512–13.
Home, Everard. "Reasons for Giving the Name Proteo-Saurus to the Fossil Skeleton Which Has Been Described." *Philosophical Transactions of the Royal Society* II (1815–1830): 113.
Hopkins, Clare. *Trinity: 450 Years of an Oxford College Community*. Oxford University Press, 2005.
Humphreys, David. *Letters from the Hon. David Humphreys to the Rt. Hon. Sir Joseph Banks Containing Some Account of the Serpent of the Ocean Frequently Seen in Gloucester Bay*. New York, 1817.
Impey, MacGregor, and Arthur Oliver. *The Origins of Museums: The Cabinet of Curiosities in Sixteenth-and Seventeenth-Century Europe*. Oxford: Oxford University Press, 2001.
Jacques-Lefèvre, Nicole. "Such an Impure, Cruel and Savage Beast." In *Werewolves, Witches and Wandering Spirits: Traditional Belief and Folklore in Early Modern Europe*, edited by Katheryn A. Edwards. Kirksville: Truman State University Press, 2002.
Janacek, Bruce. "A Virtuoso's History: Antiquarianism and the Transmission of Knowledge in the Alchemical Studies of Elias Ashmole." *Journal of the History of Ideas* 69, no. 3 (July 2008): 395–417.
Janson, H. W. *Ape and Ape Lore in the Middle Ages and Renaissance*. London: Warburg Institute, 1952.
Jardine, Lisa. *The Curious Life of Robert Hooke*. Harper Perennial, 2005.
Jerome, Joseph. *Montague Summers: A Memoir*. London: Cecil & Amelia Woolf, 1965.

J. H. H. "Review of *The Werewolf.*" *Man* 203–205 (November 1934): 182–3.
Jones, Timothy S., and David A. Sprunger, eds. *Marvels, Monsters, and Miracles: Studies in the Medieval and Early Modern Imagination.* Western Michigan University Press, 2002.
Jonston, John. *A Description of the Nature of Four-Footed Beasts.* London, 1678.
Jylkka, Katja. "Witness the Plesiosaurus: Geological Traces of the Loch Ness Monster." *Configurations* 2018, no. 2 (2018): 207–24.
Kempf, Damien, and Maria Gilbert. *Medieval Monsters.* British Library, 2015.
Kircher, Athanasius. *China Illustrata with Sacred and Secular Monuments, Various Spectacles of Nature and Art and Other Memorabilia.* Amsterdam, 1667. Translated by Charles D. Van Tuyl. Indiana University Research, 1987.
Klaniczay, Gábor, trans., Susan Singerman, ed., and Karen Margolis. *Uses of Supernatural Power: The Transformation of Popular Religion in Medieval and Early Modern Europe.* Cambridge Polity, 1990.
Knapp, Sandra. "Fact and Fantasy." *Nature* 415, no. 31 (January 2002): 479.
Knight, Gareth. *The Magical World of the Inklings.* Skylight Press, 2010.
Kong, Xurong. *Fu Poetry Along the Silk Roads: Third Century Chinese Writings on Exotica.* ARC Humanities Press, 2021.
Krag, Claus. *Sverre – Norges største middelalderkonge.* Oslo: H. Aschehoug and Co., 2005.
Kriss, Marika. *Werewolves, Shapeshifters, and Skinwalkers.* Los Angeles: Sherbourne Press, Inc., 1972.
Kruger, Lawrence. "Edward Tyson's 1860 Account of the 'Porpess' Brain and Its Place in the History of Comparative Neurology." *Journal of the History of Neurosciences* 12, no. 4 (2003): 339–49.
Lawton, John. "Nessiteras rhombopteryx." *OIKOS* 77, no. 3 (1996): 377–80.
Lee, Henry. *The Octopus: The "Devil-Fish" of Fiction and of Fact.* London: Chapman & Hall, 1875.
Lee, Henry. *Sea Monsters Unmasked.* London: William Clowes & Sons Limited, 1883.
Lee, Henry. "Sea Monsters Unmasked." In *Fisheries Exhibition Literature*, vol. III, 40–1. London: William Clowes & Sons, 1884.
Leith Ross, Prudence. *The John Tradescants: Gardeners to the Rose and Lily Queen.* Peter Owen Publishers, 2006.
Ley, Willy. "Scylla Was a Squid." *Natural History Magazine*, June 1941.
Ley, Willy. "Do Prehistoric Monsters Still Exist?" *Mechanix Illustrated* (February 1949): 80–144.
Ley, Willy. *Exotic Zoology.* New York: Viking Press, 1955.
Liceti, Fortunio. *De Monstris.* Amsterdam, 1665.
The Linnean Society of London. *An Essay on the Credibility of the Existence of the Kraken, Sea-Serpent, and Other Sea Monsters: With Illustrations.* London: William Tegg, 1849.
Llewellyn-Jones, Lloyd. *Ctesias' 'History of Persia': Tales of the Orient.* Routledge, 2012.
Lord, Percival. "Review of Hawkins' *Memoirs.*" *Athenaeum* 347 (1834): 469–70.
Loxton, Daniel, and Donald R. Prothero. *Abominable Science: Origins of the Yeti, Nessie, and other Famous Cryptids.* Columbia University Press, 2015.
Lucretius. *On the Nature of Things.* Translated by Martin Ferguson Smith. Hackett, 2001.
MacKal, Roy P. *A Living Dinosaur?: In Search of Mokele-Mbembe.* New York: E. J. Brill, 1987.
Mamigliano, Arnaldo. "Ancient History and the Antiquarian." *Journal of the Warburg and Courtauld Institute* 13 (1950): 285–315.
Marks, Jonathan. "Contemporary Bio-Anthropology: Where the Trailing Edge of Anthropology Meets the Leading Edge of Bioethics." *Anthropology Today* 18, no. 4 (August 2002): 3–7.

Mauries, Patrick. *Cabinets of Curiosities*. Thams and Hudson, 2002.
McDermott, William. *The Ape in Antiquity*. Johns Hopkins University Studies in Archaeology 27, 1938.
McGowan, C. "Problematic Ichthyosaurs from Southwest England: A Question of Authenticity." *Journal of Vertebrate Paleontology* 10, no. 1 (March 1990): 72–9.
Megasthenes. *Indica*. Routledge, 2023.
Meldrum, Jeff. *Sasquatch: Legend Meets Science*. Forge Trade, 2009.
Mellick, S. A. "Dr Nicolaes Tulp of Amsterdam, 1593–1674: Anatomist and Doctor of Medicine." *Australian and New Zealand Journal of Surgery* 77, no. 12 (December 2007): 1102–9.
Melville, Herman. *Moby Dick: Or the Whale*. Harper and Brothers, 1851.
Montagu, Ashley. *Edward Tyson, M.D., F.R.S., 1650–1708*. Philadelphia: American Philosophical Society, 1943.
Moran III, Francis. "Of Pongos and Men: 'Orang-Outang' in Rousseau's 'Discourse on Inequality'." *Review of Politics* 57, no. 4 (Autumn 1995): 641–64.
Moseley, C. W. R. D., trans. *The Travels of Sir John Mandeville*. London: Penguin Books, 1983.
M'Quhae, Captain Peter. "Response to Professor Owen." *Zoologist* 6 (1848): 2323–4.
Newenfeldt, Georg. *Disquisitio Philosophica, Vol I De Oraculis Gentilium, Vol II De Lycanthropia*. Jenae, 1650.
"The New England Sea-Serpent." *Western Folklore* 7, no. 1 (January 1948): 67.
Newman, Charles. "Edward Tyson." *British Medical Journal* (October 11, 1975): 96–7.
Nickell, Joe. "The Skeptic-Raping Demon of Zanzibar." *Skeptical Briefs*, December 1995.
Nordoff, Charles. "A Plea for a Monster." *Harper's New Monthly Magazine* XXI (July 1860): 178–93.
O'Connell, Sheila, and David Paisey. "This Horryble Monster: An Anglo-German Broadside of 1531." *Print Quarterly* 16, no. 1 (March 1999).
O'Connor, Ralph. *The Earth on Show: Fossils and the Poetics of Popular Science, 1802–1856*. Chicago: University of Chicago Press, 2007.
O'Donnell, Elliot. *Werwolves*. London: Methuen & Co., 1912.
O'Donnell, Elliot. *Twenty Years as a Ghost Hunter*. London: Heath, Cranton Ltd., 1917.
"On the History of the Great Sea-serpent." *Blackwood's Edinburgh Magazine* XIII, no. III (April 1818): 32–42.
O'Neill, J. P. *The Great New England Sea-Serpent*. Camden, ME: Down East Books, 1999.
Otten, Charlotte. *A Lycanthropy Reader*. Syracuse University Press, 1986, 5–8.
Oudemans, A. C. *The Great Sea-Serpent*. Leiden: E.J. Brill; London: Luzac & Co., 1892.
Ovid. *Metamorphoses*. Translated by Bernard Knox. W.W. Norton, 2004.
Owen, Richard. "Notebook: Collection of Newspaper Cuttings Relating to the Alleged Appearance of the 'Sea-Serpent'." OC36, Owen Papers, Library, Natural History Museum, London.
Owen, Richard. *The Life of Richard Owen*. London: John Murray, 1894, 334.
Packard, Jr. A. S. "Colossal Cuttlefishes." *American Naturalist* 7, no. 2 (February 1873): 87–94.
Park, Katherine, and Lorraine Daston. "Unnatural Conceptions: The Study of Monsters in Sixteenth and Seventeenth Century France and England." *Past & Present* 92 (August 1981).
Patterson, Serina. "Reading the Medieval in Early Modern Monster Culture." *Studies in Philology* 111, no. 2 (Spring 2014).
Paxton, C. G. M., E. Knatterud, and S. L. Hedley. "Cetaceans, Sex and Sea Serpents: An Analysis of the Egede Accounts of a 'Most Dreadful Monster' Seen Off the Coast of Greenland in 1734." *Archives of Natural History* 32, no. 1 (2005): 1–9.

Pliny. *Natural History*, vol. 3. Loeb Library, 1940, 59. Also see, Richard Henry Crum. "The Werewolf Again." *Classical Weekly* xxvi, no. 13 (January 23, 1933).
Pliny's Natural History in Thirty Seven Books. Lagare Street Press, 2022.
Porshnev, Boris. *The Soviet Sasquatch*. UK: CFZ Press, 2021.
Prieur, Claude. *Dialogue de la lycanthropie, ou, Transformation d'hommes en loups*. Louuain, 1596.
Pumfrey, Stephen. "Who Did the Work? Experimental Philosophers and Public Demonstrators in Augustan England." *British Journal for the History of Science* 28 (1995): 131–56.
Radford, Benjamin. "Popobawa! In Search of Zanzibar's Bat-Winged Terror." *Fortean Times* 241 (November 2008): 34–9.
Rafinesque, C. S. "Dissertation on Water Snakes, Sea Snakes, and Sea Serpents." *American Monthly Magazine and Critical Review* 1 (October 1817): 431–5.
Redfern, Nick. *Man-Monkey: In Search of the British Bigfoot*. CFZ Press, 2007.
Redfield, James. "Herodotus the Tourist." *Classical Philology* 80, no. 2 (April 1985): 97–118.
Regal, Brian. "The Occult Life of Montague Summers." *Fortean Times* 349 (February 2017): 42–6.
Richards, Sabrina. "The World in a Cabinet, 1600s." *The Scientist* 26, no. 4 (April 1, 2012).
Ritvo, Harriet. "Professional Scientists and Amateur Mermaids: Beating the Bounds in Nineteenth-Century Britain." *Victorian Literature and Culture* (1991): 277–91.
Robinson, Harry W., and Walter Adams, eds. *The Diary of Robert Hooke, 1627–1680*. London, 1935, 293.
Robinson, Joyce Henry. "An American Cabinet of Curiosities: Thomas Jeffersons Indian Hall at Monticello." *Winterthur Portfolio* 30, no. 1 (Spring 1995).
Romero-Reverón, Rafael, and Luis A. Arráez-Aybar. "Ole Worm (1588–1654) – Anatomist and Antiquarian." *European Journal of Anatomy* 19, no. 3 (2015).
Ross, Sydney. "Scientist: The Story of a Word." *Annals of Science* 18, no. 2 (June 1962): 65–85.
Rudwick, Martin. *Scenes from Deep Time: Early Pictorial Representations of the Prehistoric World*. University of Chicago Press, 1995.
Rupke, Nicolaas A. *Richard Owen: Victorian Naturalist*. Yale University Press, 1994.
Rupke, Nicolaas A. *Richard Owen: Biology without Darwin*. Rev. ed. Chicago: University of Chicago Press, 2009.
Saint-Hilaire, Isidore Geoffroy. *Histoire générale et particulière des anomalies de l'organisation chez l'homme et les animaux*, 4 vols. Paris: J.B. Bailliere, 1832–1837.
Sanderson, Ivan. *Abominable Snowmen: Legend Come to Life*. Chilton, 1961.
Sanderson, Ivan. "Preliminary Description of the External Morphology of What Appeared to Be the Fresh Corpse of a Hitherto Unknown form of Living Hominid." *Genus* 25 (1969): 259–78.
Schotanus, M. Susanne. *Interdisciplinary Essays on Monsters and the Monstrous*. Bingley: Emerald Publishing, 2022.
"The Sea-Serpent Again." *Standard*, September 18, 1877.
Secord, James. "Monsters at the Crystal Palace." In *Models: The Third Dimension of Science*, edited by S. de Chadarevian and N. Hopwood. Stanford University Press, 2004.
Semonin, P. *American Monster: How the Nation's First Prehistoric Creature Became a Symbol of National Identity*. New York: NYU Press, 2000.
Senn, Harry. "Romanian Werewolves: Seasons, Ritual, Cycles." *Folklore* 93, no. 2 (1983): 206–15.
Shuker, Karl. "From Star Wars to Smithfield, the Curious Case of the Medieval Yoda Monk." *Shuker Nature*, December 17, 2015.

Smith, J. E. H. "Language, Bipedalism, and the Mind-Body Problem in Edward Tyson's Orang-Outang (1699)." *Intellectual History Review* 17, no. 3 (2007): 291–304.

Smith, J. L. B. *Old Fourlegs: The Story of the Coelacanth*. Longmans & Green, 1956.

Smith, Jay M. *Monsters of the Gévaudan: The Making of a Beast*. Harvard University Press, 2011.

Sterling, Keir B., ed. *Rafinesque: Autobiography and Lives*. New York: Arno Press, 1978.

Strothers, Richard B. "Ancient Scientific Basis of the 'Great Serpent' from Historical Evidence Evidence." *Isis* 95 (2004): 220–38.

Struit, Dirk Jan. *Yankee Science in the Making: Science and Engineering in New England from Colonial Times to the Civil War*. New York: Dover Publications, 1991.

Summers, Montague. *The Werewolf*. London: Kegan Paul, 1933.

Summers, Montague, ed. *The Supernatural Omnibus*. London: Victor Gollancz Ltd., 1974.

Summers, Montague. *The Galanty Show*. London: C. Woolf, 1980.

Sutin, Lawrence. *Do What Thou Wilt: A Life of Aleister Crowley*. New York: St. Martin's Griffin, 2000, 338–43.

Sweeney, James B. *A Pictorial History of Sea Monsters and Other dangerous Marine Life*. New York: Crown Publishing, 1972.

Tatham, David. "Elihu Vedder's *Lair of the Sea-serpent*." *American Art Journal* 12, no. 2 (Spring 1985): 33–47.

Thom's Irish Who's Who: A Biographical Reference of Prominent Men and Women in Irish Life at Home and Abroad. Alexander Thom &n Co., 1923.

Thwaite, Ann. *Glimpses of the Wonderful: The Life of Philip Henry Gosse*. Faber & Faber, 2002.

Tulp, Nicholaas. *Observationes Medicae*, 1641.

Tyson, Edward. *Phocaena, or the Anatomy of the Porpess Dissected at Gresham College*. London, 1680.

Tyson, Edward. "An Observation of an Infant Where the Brain Was Depressed into the Hollow of the Vertebrae of the Neck." *Philosophical Transactions of the Royal Society* 19 (1695–1697): 533–5.

Tyson, Edward. "A Relation of Two Monstrous Pigs with the Resemblance of Human Faces." *Philosophical Transactions of the Royal Society* 21 (1699): 431–5.

Ucko, Peter. "Subjectivity and the Recording of Palaeolithic Cave Art." In *The Limitations of Archaeological Knowledge*, edited by T. Shay and J. Clottes, 141–80. Liege: University of Liege Press, 1992.

Van Rys, John C. "Reclaiming Marginalia: The Grotesque and the Christian Reader." *Christianity and Literature* 44, no. 3/4 (Spring–Summer 1995).

Verrill, A. E. "The Colossal Cephalopods of the North Atlantic II." *American Naturalist* 9, no. 2 (February 1875): 78–86.

Verstegen, Richard. *A Restitution of Decayed Intelligence*. London: John Norton, 1634.

Voyage La Guyane française: notes et souvenirs d'un voyage exécuté en 1862–1863. Paris: Hachette, 1867.

Warren, Leonard. *Constantine Samuel Rafinesque: A Voice in the American Wilderness*. University of Kentucky Press, 2004, 164.

Weinreich, Spencer J. "Thinking with Crocodiles: An Iconic Animal at the Intersection of Early-Modern Religion and Natural Philosophy." *Early Science and Medicine* 20–3 (2015).

Westrum, Ron. "Knowledge About Sea-Serpents." In *On the Margins of Science: The Social Construction of Rejected Knowledge*, edited by Roy Wallis. University of Keele, 1979.

Whitelock, Dorothy, ed. *English Historical Documents: 500–1042*. Routledge, 1979.
Wickham, D. E., ed. *Letters to an Editor: Montague Summers to C.K. Ogden*. Edinburgh: Tragara Press, 1986.
Williams, David. *Deformed Discourse: The Function of the Monster in Medieval Thought and Literature*. McGill-Queen's University Press, 1996.
Wilson, A. N. *C. S. Lewis: A Biography*. New York: W.W. Norton, 1990.
Wittkower, Rudolf. "Marvels of the East: A Study in the History of Monsters." *Journal of the Warburg and Courtauld Institutes* 5 (1942): 159–97.
Wood, J. G. "Trial of the Sea-Serpent." *Atlantic Monthly* 53 (1884): 799–814.
Worm, Ole. *Museum Wormianum: Historia Rerum Rariorum*. Lugduni Batavorum, 1655.

INDEX

Note: Page numbers followed by an 'n' indicate a note on the corresponding page.

For Product Safety Concerns and Information please contact our EU representative GPSR@taylorandfrancis.com Taylor & Francis Verlag GmbH, Kaufingerstraße 24, 80331 München, Germany